Handbook of tex

Principles, processes

Jacquie Wilson

The Textile Institute

D1089393

CRC Press

Boca Raton Boston New York Washington, DC

WOODHEAD PUBLISHING LIMITED

Cambridge England

Published by Woodhead Publishing Limited in association with The Textile Institute
Abington Hall, Abington
Cambridge CB1 6AH
England
www.woodhead-publishing.com

Published in North and South America by CRC Press LLC,
2000 Corporate Blvd, NW Boca Raton FL 33431, USA

Ref
TS
1475
.W55
2001

First published 2001, Woodhead Publishing Ltd and CRC Press LLC
© Woodhead Publishing Ltd, 2001
The author has asserted her moral rights.

British Library Cataloguing in Publication Data
A catalogue record for this book is available from the British Library.

Library of Congress Cataloging in Publication Data
A catalog record for this book is available from the Library of Congress.

Woodhead Publishing ISBN 1 85573 573 3
CRC Press ISBN 0-8493-1312-0
CRC Press order number: WP1312

Cover design by the ColourStudio
Typeset by Replika Press Pvt Ltd, Delhi 110 040, India
Printed by TJ International, Cornwall, England

Contents

This book is dedicated to the memory of my mum who would have been a terrific textile designer if she had had the opportunity, and to my kids Flynn and Blue for their patience (most of the time) with a working mum.

Preface

Since being a student in Galashiels in the early 1970s I have felt there has been a need for some sort of text that covered the textile design process from initial ideas through briefing, research and design development, to finished fabrics being sold to garment designers and to retail. What follows is an attempt to provide such a text. This book is based on my experiences as a textile designer in industry and my teaching at UMIST.

With the other commitments in my life it has taken longer to write this book than I had initially anticipated. The more I have researched and written, the more I have come to realise that there is so much more that could be included. However, there came a point where I felt I had to follow my own advice that I had enough information for the project to stop, and to get on with putting a final manuscript together.

I am aware that there are gaps; for example there is nothing on carpet design or warp knitting, and little on CAD systems. I would like to think that a second edition will address some of the gaps such as the carpet design and warp knitting; however, with regard to CAD, I wanted to concentrate more on processes and felt that to include a lot about CAD systems would not be particularly helpful and would become very quickly dated.

Anyway, here it is, and while not everything is covered I think it does fill a gap and should be of value to students of textile design around the world.

Jacquie Wilson

Acknowledgements

Many people have had an influence on the contents of this book, either directly or indirectly, and I would like to take this opportunity to thank them.

Thanks to all those at the Scottish College of Textiles who were there in the early 70s who provided me with inspiration then and throughout my career. Special mention must go to Leslie Blythe, Tom Stillie, Ian Mackenzie Gray, Dr Martindale and Sandy Cass who are sadly no longer with us, and to Ronnie Moore and Leslie Millar who I hope are enjoying well-deserved retirement. Acknowledgement must also go to my fellow students, particularly Ron Hall whose work was always so good we all had to work extra hard to try to keep up!

Thanks also to all those I have worked with during my time in industry and all those I have worked with since I came to UMIST in 1984.

I must also mention all those students who have sat through the lectures that have formed much of the basis for this book — I hope I have been of some help to them.

A big thank you goes to Patricia Morrison at Woodhead Publishing for believing in this book, particularly for her support and patience over the last four years.

1

An overview of textiles and textile design from fibre to product purchase

1.1 The global textile and clothing industries

Textile making is a very ancient craft, with a history almost as old as mankind itself. Remembered and recorded in poetry and ancient stories and myths, textiles have always been important to man. As well as providing protection from the elements, the first textiles were used as decoration, providing status for the owner. They were also used as tools; bags for transporting belongings and for holding food as it was gathered.

Textiles are produced in almost every country of the world, sometimes for consumption exclusively in the country of manufacture, sometimes mainly for export. From cottage industry to multi-national corporation, textiles and clothing are truly global industries.

In 1782, the invention of the steam engine gave the world a new power source and started the Industrial Revolution. Previous to this the production of textiles had been a domestic system, a cottage industry with textiles spun, knitted and woven in the home. By the middle of the nineteenth century, however, there was a whole range of new machines and inventions that were to take textiles into an era of mass production in factories. The development of man-made fibres and new dyestuffs in the early part of the twentieth century, and continuing technological developments, have led and continue to lead to new products and applications. The actual processes of textile manufacture, however, are still very much as they have always been, with the vast majority of cloth being woven or knitted from yarn spun from fibre. And, while much production may be very technologically advanced, hand-produced textiles are still made in many countries exactly as they were many, many years ago.

Nowadays, many different types of companies are involved in the production of textiles and clothing world-wide; some companies own many huge manufacturing plants in many different countries while others will have only a few employees and some may not actually manufacture at all.

1.2 Textile materials, processes, and products

Fibres are manufactured or processed into yarns, and yarns are made into fabrics. Fabrics may be manufactured by a variety of processes including knitting, weaving, lace-making, felt-making, knotting (as in some rug and carpet manufacture), and stitch bonding. These fabrics may be industrial textiles with detailed technical and performance specifications, or they may be sold either to retail or contract as apparel, furnishings or household textiles, where aesthetics may be as, or sometimes even more important than performance. The fabrics may be coloured by dyeing or printing, or be finished to enhance their appearance (such as by brushing) or performance (such as by application of a flame-retardant). A wide diversity of products are made from textile products or have some textile components; textiles go into car tyres, and geotextiles are used for lining reservoirs, while medical applications include artificial ligaments and replacement arteries. Figure 1.1 summarises textile materials, processes and products in chart form.

1.2.1   Design in textiles and clothing

Every textile product is designed: that is, it is made specifically to some kind of plan. Design decisions are made at every stage in the manufacturing process — what fibres should be used in a yarn, what yarns in a fabric, what weight of fabric should be produced, what colours should the yarn or fabric be produced in, what fabric structures should be used and what finishes applied. These decisions may be made by engineers and technologists in the case of industrial or medical textiles where performance requirements are paramount, or, more often in the case of apparel, furnishings and household textiles, by designers trained in aesthetics, technology and marketing. The designers found in the textile and clothing industries are frequently involved throughout the design process, from initial identification of a need/requirement, through research, generation of initial design ideas, design development and testing to ultimate product specification.

1.2.2   Designers found in the textiles and clothing industries

The designers found in textiles and clothing include:

- colourists predicting and forecasting future colour ranges
- yarn designers
- knitted fabric designers
- woven fabric designers
- carpet designers
- print designers
- embroidery designers
- knitwear designers
- garment designers
- accessory designers
- print producers
- stylists
- colourists developing colourways
- repeat artists

1.2.3   Fibres

Fabric is made from yarn, and yarn is made from fibres. These fibres can be either natural or man-made.

Natural fibres include animal fibres (e.g. wool and silk), vegetable fibres (e.g. jute and cotton) and mineral fibres (e.g. asbestos). Man-made fibres are either regenerated or synthetic; viscose rayon, based on regenerated cellulose, is man-made but not synthetic while polyester, polypropylene and nylon are all synthetic fibres.

Synthetic fibres are produced by the large chemical companies including Dupont, Bayer, Hoechst and Astra Zeneca. Many of these companies produce no fabric but specialise in the production of certain types of fibre which they sell on as fibres or manufacture into yarns.

1.2.4 Yarns

Yarn producers or spinners buy in natural and/or man-made fibres to make these into yarns of different sizes and characters; regular and fancy yarns. For many years the main spinning systems could be given as woollen, worsted and cotton, and these systems gave rise to the woollen, worsted and cotton industries. Developments in spinning, however, have led to new spinning systems including 'open-end', 'self-twist' and 'jet' spinning.

At its simplest, yarn production is essentially about taking fibres, organising them so that they lie in a lengthways direction and twisting them to create a yarn. By combining fibre types, and using different spinning systems and machinery, yarns can be developed with individual profiles suitable for a vast range of end uses. Regular yarns are those which have a regular straight profile and these can be twisted together, making 'two-fold' or 'three-fold' yarns for example. Fancy yarns can be created by deliberately introducing irregularities or intermittent effects along their length. Yarns can be combined together as components of new yarns with different effects and properties from their component parts. As well as changing the appearance of a fabric, the introduction of a fancy yarn will affect the handle and performance of that fabric.

1.2.5 Woven fabrics

Strictly speaking, the definition of a textile is 'a woven fabric' but the term textile is now considered to cover any product that uses textile materials or is made by textile processes.

Essentially, woven fabrics are structures produced by interlacing two sets of threads; the warp which runs in a lengthways direction and the weft which runs in a widthways direction. Weaving methods include tapestry and jacquard.

1.2.6 Knitted fabrics

Knitted fabrics are produced by interlacing loops of yarn. In weft knitting, loops are formed one at a time in a weft-ways direction as the fabric is formed. Hand-knitting with a pair of knitting needles is weft knitting. In warp knitting there is a set of warp yarns which are simultaneously formed into loops. To connect these chains of loops the warp threads are moved sideways in such a way as to cause the loops to interlink.

1.2.7 Lace and non-woven fabrics

Fabrics may also be produced by methods other than weaving and knitting. Lace is an open-work fabric made by looping, plaiting or twisting threads by means of a needle or a set of bobbins. Fabrics produced by crochet and macramé are often called lace, although strictly speaking they are not. Knotting is another way of making fabrics. Knotting was a popular pastime for women in eighteenth-century Europe and colonial

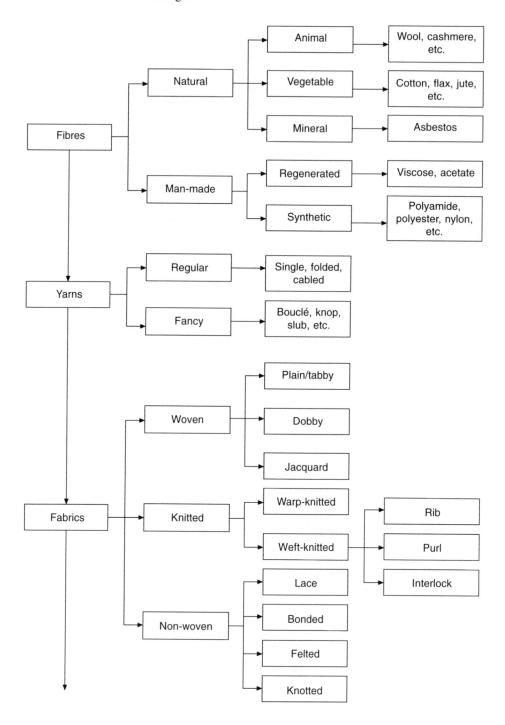

Fig. 1.1   Textile materials, processes and products.

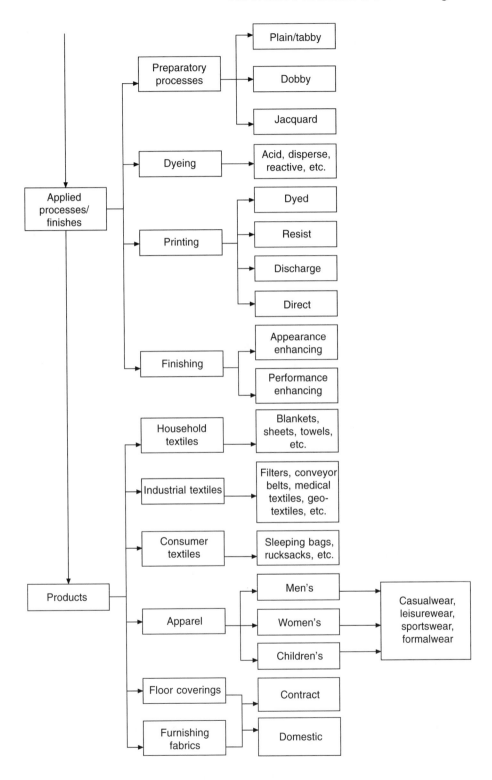

Fig. 1.1 (*Contd*)

North America, and one method still seen today is macramé. A knotting process is also used for fishing nets, and some rugs and carpets are knotted — made by tying yarns onto a foundation weave.

There is also a group of fabrics called non-wovens which include true felt (where animal fibres are matted together) and fabrics produced by bonding webs of fibres together by stitching or by sticking with adhesive. However, in terms of volume produced, knitted and woven fabrics are by far the most common methods of fabric production.

1.2.8   Fabric terms

A length of woven or knitted fabric is usually referred to as a 'piece'. Often, fabric woven by a mill will not be coloured and this undyed fabric is called 'grey cloth'. Colour can be added by dyeing the piece, and such fabric is referred to as being 'piece-dyed'. Colour can also be added to a fabric by applying pigments or dyes in a printing or other colouring process after weaving or knitting, or by using already dyed yarns in the construction of the fabric. Cloth made from dyed yarns will not normally be dyed again or printed.

'Finishing' is what happens after the fabric has been made. The finishing processes employed will be determined by the type of fabric and its performance requirements. Any excess dye will normally be removed, any applied pigment will normally be set, and any dye will be fixed. Fabrics may be brushed or raised to enhance appearance and handle, or fire-retardant and soil-resist treatments may be applied. Fire retardancy may be a product performance prerequisite; anti-soiling and anti-static finishes, while not necessarily pre-requisites, enhance performance, as do methods of coating fabrics to produce microporous surfaces.

1.2.9   Geography and fabric types

Certain countries, and areas within countries, have developed industries around specific fibres/fabric types and there are still parts of the world where craftsmen produce fabrics exclusive to them, such as some hand-crafted batiks and weaves.

In the UK, Manchester was nicknamed 'Cottonopolis' as it was built in the main (as was much of Lancashire) from money earned through the cotton trade. Scotland developed a woollen trade through both woven and knitted fabrics, while, again in the UK, Yorkshire was home to the worsted industry and the Midlands became famous for knitting and lace making. India has a history of cotton manufacture and, in the eighteenth century, was famous throughout Europe for its mordanted cottons or chintzes.

1.3   Textile organisations

1.3.1   Size and structure

The factories producing textiles are usually called mills. Some mills take in fibre, spin yarns, dye these and then either weave or knit these into fabrics/garments. Such organisations that are involved in several textile processes are described as 'vertical'. There are, however, also plenty of companies (often smaller but not always) specialising in one of these functions, usually on a commission basis, and these organisations are described as 'horizontal'. There are many examples of commission knitters, printers, dyers and finishers — companies that for a unit price process textile goods for other companies in the industry.

Many textile organisations today are huge multi-national corporations involved in more than one textile process; producing fibres, spinning, dyeing, weaving and knitting, printing, and garment manufacture. These companies will often have these various processes carried out in many different countries. It is therefore not unusual to buy a garment in Japan or the USA that was made up in Portugal, with sewing threads from the UK, from fabric woven in Korea, from yarn manufactured in Italy and from fibres made in Germany.

Until the 1960s and 70s, most textile companies in the UK, Europe and the USA were relatively small organisations. Many were vertical operations, involved in all the manufacturing processes from fibre to finished product, although there were some horizontal organisations, specialising in only one process such as spinning, weaving, or dyeing and printing. There was a period of major change in the 1970s when many mergers and take-overs took place resulting in re-groupings of operations. Textiles and clothing in the twenty-first century will continue to be a truly global industry.

1.3.2 Converters and wholesalers

Manufacturers of grey cloth may sell this fabric to converters rather than do anything further to it themselves. Converters buy grey cloth and convert this by having it dyed or printed, and then finished. A mill will own specific equipment, or plant, which must be kept operating to maintain profitability; a converter has greater flexibility in that such an operation does not need to own any equipment, having everything done by other organisations. If a converting company has a new idea, they can find a new resource without compromising existing business. A mill, however, with all its operations under one roof, does have more control.

As we have seen, fabrics are usually produced by the piece. Mills and converters usually sell by the piece. Wholesalers essentially buy from a manufacturer and, without changing the product, sell it in smaller quantities to retailers or smaller manufacturers.

1.4 Categorising textiles

Textiles can be categorised by production company, by end-use, and by the market for which they are designed.

1.4.1 Production companies

The companies involved in textile and clothing production can be grouped in different ways, for example by product or by manufacturing process. Many companies fall into several categories. A mill may produce fabrics but it may also convert fabric that it is uneconomical for it to produce itself. Usually labelled by the function which they primarily perform, i.e. a spinner, a knitter, a weaver, etc., companies may also be known for the end-use of their fabric, i.e. as a producer of men's knitwear, lingerie, contract furnishings, etc. Mills often produce fabrics for many different end-uses while converters mostly develop fabrics exclusively for one end-use.

A company might describe itself as: 'Hosiery manufacturers, and spinners and doublers of super merinos in white and colour. Also manufacturers of high quality underwear and knitwear for men and women.'

1.4.2   End-uses

Textiles are found in a hugely diverse range of products. Clothing us from birth until death, textiles protect us and make us feel good. Our homes are made more comfortable by textiles that keep in heat and by textiles that shield us from the sun. Keeping us warm at night, textiles also dry us when we are wet and can support injured limbs. Textiles allow us to make tea directly in a cup. More recently specialised textiles have been developed for medical use as artificial replacement ligaments and arteries, and geotextiles are used in the construction of dams and motorways and even bunkers on golf courses.

1.4.2.1   Apparel textiles

The clothing or apparel market includes most garments that are worn. A huge consumer of fabric, clothing manufacture can be split by market, e.g. men's, women's and children's clothing, sportswear, casual wear or formal wear. However, not all fabrics for garments are considered part of the apparel market. Fabrics such as the specialised protective clothing for fire-fighters, pilots and those in similar hazardous occupations are considered part of the industrial textiles market, and specialist clothing for leisure and ski wear, etc. are considered as being consumer textiles.

1.4.2.2   Furnishing fabrics or interior textiles

The furnishing market is another huge consumer of textiles, for curtains, upholstery fabrics, carpets and wall coverings, either domestic or contract. Domestic furnishings are those found in the home, while contract furnishings are those used in offices and public buildings such as schools, hotels and hospitals.

1.4.2.3   Household textiles

This category includes all textile products used within the home except furnishings, including sheets, pillowcases, towels, blankets, tablecloths, etc. When these products are used in the contract market they may be referred to as 'institutional fabrics'.

1.4.2.4   Industrial textiles

Car tyres, medical textiles and geotextiles are all examples of industrial textiles. Industrial textiles also covers such textile products as filters, conveyor belts, car safety belts and parachute cords. Performance is of prime importance in this category.

1.4.2.5   Consumer textiles

This category could be described as including any textiles not falling into the previous categories. Recreational items such as tents and back packs may be referred to as consumer textiles, as well as awnings and umbrellas and luggage. Although in this category performance can be very important, aesthetics can be equally so.

1.4.3   Textiles categorised by market area and price

Textiles and textile products can be categorised by the market area for which these are intended and by price. Expensive fabrics, apparel and furnishing products may be described as 'upper end', 'top', 'exclusive', 'haute couture' and 'designer'. However, the largest quantity of fabrics and textile products are sold in the middle volume, or mass market area, and in the lower, down-market area.

1.4.4 Categorising textile companies

Any textile company can be described using various labels — by the manufacturing process carried out, by the product type and by product market area and price. Companies may also be known as volume converters, top-end fashion-fabric producers, cut and sewn or fully-fashioned knitters, etc.

1.5 Summary

The textiles and clothing industry is a large and diverse global industry. While technology has had a tremendous impact on some aspects of textile production, there are other areas where processes have changed little from those first developed. There are many designers employed in this industry, in a wide variety of different positions and with a wide range of roles and responsibilities. Textile organisations can be large or small and can be classed in a variety of different ways. Mills may be vertical or horizontal. The true globalisation of the textile and clothing industry will continue to develop in the twenty-first century.

Bibliography

Corbman, B.P., *Textiles: Fibre to Fabric*, 6th ed., New York; London, Gregg, 1983.

McIntyre, J.E. and Daniels, P.N. (eds), *Textile Terms and Definitions*, 10th ed., Manchester, Textile Institute, 1995.

Taylor, M.A., *Technology of Textile Properties: An Introduction*, 3rd ed., London, Forbes Publications, 1990.

Totora, P.G. and Collier, B.J., *Understanding Textiles*, 5th ed., London, Prentice Hall, 2001.

Yates, M., *Textiles: A Handbook for Designers*, rev. ed., New York; London, W.W. Norton, 1996.

2

Textile designers

2.1 The diversity of textile design and textile designers

The diversity of the textile and clothing industries is reflected in the many different types of designers needed. At every stage of manufacture of textiles there are colourists determining the fashion colours in which the fibres will be produced, yarn designers developing yarns to meet certain requirements, knitted-fabric designers, woven-fabric designers, carpet designers, print designers, embroidery designers, knitwear designers, designers of women's wear, men's wear and children's wear, accessory designers, designers of casualwear, sportswear, eveningwear, swimwear and designers for the mass market, haute couture and designer labels, etc.

2.1.1 The purpose of the textile designer

The role of the designer can be quite complex but the overall purpose can be stated quite simply — the textile designer has to design and produce, to an agreed timetable, an agreed number of commercially viable fabric designs. Depending on the markets that he or she is designing for, several different activities are involved in fabric design and the number and type in which any designer is involved will vary according to the product and production methods used, and the type of company for which the work is done.

2.1.2 Stylists

Designers also put together ranges. For example a stylist might handle the development of a company's range of printed fabrics. A range is a group of fabrics (or products) designed, developed and edited to be shown and sold to the market each season. The stylist initiates the design work, organises and directs the development and coloration of intended designs (frequently using freelance designers), and co-ordinates with manufacturing personnel to have samples made. These samples are shown to customers; the stylist then edits and finalises the group of designs that will form that season's range.

Further down the chain, buyers and merchandisers in retail organisations do much the same range-building processes.

2.1.3   Colourists

Some designers work purely with colour, predicting colour trends and putting together palettes of colours for specific seasons and product groups. Other colourists will work further down the design process line, colouring designs produced by other designers to create different and alternative colourways.

2.1.4   Repeat artists

A company producing printed textiles will often employ designers whose main function is to take designs and put these into a size and repeat appropriate to the intended end-use.

2.1.5   In-house and freelance designers

Designers may work for manufacturing companies as in-house designers, or they may work independently as freelance designers. In-house designers, or as they are sometimes called staff designers, are employed by a company usually on a full-time basis, although some may be employed part-time. Often they work within a manufacturing environment, although they can also be employed by retailers and by converters.

Freelance designers may either work for independent studios or through an agent, producing designs on paper for which the studio/agent receives a commission when the designs are sold to mills and converters. Alternatively, freelance designers may put together a portfolio of their designs, which they may sell directly to stylists. The work they produce for their portfolio, while it will have at least to reflect trends, will be often very much what they themselves like and want to produce. While the designer may have a view of the type of customer who will buy their designs when the design is developed, there may be no specific customer waiting to buy their work on completion.

Freelance designers may also develop design work according to a stylist's specification. For example, a freelance print designer who is particularly good at intricate florals may well be approached by a stylist to work on a specific print idea that will form part of that company's new season's collection. The brief may include size details, colouring details and even the type of flowers to be painted. This work will be commissioned in advance. The designer develops their paperwork with the knowledge that when it is finished there is a buyer for it.

Freelance weave and knit designers will normally work on a specific project with a manufacturer. They will be commissioned to produce a range of fabrics, or, in the case of a knitwear designer, a range of knitted-garment designs.

A third type of designer found within textiles is the consultant designer. A consultant is employed by a company to advise on design matters and may be given the task of managing the design programme. A consultant designer will usually work for several companies at any one time, although their contract may be such as to impose restrictions on their working for closely related organisations. Very often consultants will do little actual working-through of design ideas themselves; rather they will make design policy decisions and direct other designers who may be in-house or freelance.

All designer systems have advantages and disadvantages to the designers themselves and to the organisations for whom they work. These are summarised in Tables 2.1–2.3.

Just as many different people doing different jobs are given the name designer, so too are many rooms called design studios. The 'studio' can be anything from an area set aside on the factory floor to a large, smart office or even suite of offices. One designer

Table 2.1   Advantages/disadvantages of designer systems — in-house designer.

In-house designer

Advantages	**Disadvantages**
to the company —	*to the company* —
Knows	
product	Can become stale
production processes	Not always right for job
customers	
Exclusive design work	
Always there	
Can be used for jobs other	
than pure design jobs	
to the designer —	*to the designer* —
Familiar with	Can become bored
product	Lack of challenge
processes	Not always right for job
customers	May find designing for one company
Paid to travel — expenses	limiting
Known regular income	Can't earn more
May be used for jobs other than design jobs	

Table 2.2   Advantages/disadvantages of designer systems — freelance designer.

Freelance designer

Advantages	**Disadvantages**
to the company —	*to the company* —
Fresh input of ideas	Doesn't know
Only paid for design work	product
as and when required	production
Can use most appropriate	processes
designer for job	customers
Gets out and about — not at	Design work not exclusive
company's expense	Not always there when required
to the designer —	*to the designer* —
Doesn't get bored	Not familiar with
New challenges	product
variety of projects,	process
can organise own work	customers
schedule	Busy times of year when all clients want
Can visit shows/exhibitions	attention at the same time
	Has to pay for own travel
	May have to give guarantee not to work for
	competitor of a client

or a whole team of designers can work in a studio and a few or many of the functions of, and stages in, the design process might be carried out there.

2.2   Timing in the textile and clothing industries

Products are planned and produced well ahead of retail selling seasons in all areas of the textile and clothing industries. In August, and even July, autumn clothing appears

Table 2.3   Advantages/disadvantages of designer systems — consultant designer.

Consultant designer	
Advantages	**Disadvantages**
to the company —	*to the company* —
Expertise not within the company	Can be expensive
can be bought in	Not always there when required
Can sell to customers	May have split loyalties
Manages design programme	
Works closely with in-house design team	
to the designer —	*to the designer* —
Knows income	May have split loyalties
	Usually on a retainer, so often called in when inconvenient

in the shops. Before this clothing can be designed, fabrics must be designed and shown to clothing manufacturers, and orders placed, produced and delivered. Before then, new yarns and sometimes even new fibres must be developed. The scheduling and amount of time necessary for all of these steps are dependent on the amount of change occurring in the product, the volume being produced and the efficiency of the producing companies. At a minimum, textiles are usually designed a year to a year-and-a-half in front of the retail season. Major changes such as the development of a completely new type of fabric will take even longer.

2.2.1   Seasonal ranges
Manufacturers producing for the fashion industry will normally launch two main ranges a year; for the selling seasons Spring/Summer and Autumn/Winter. Sometimes, mid-season ranges are also produced. Manufacturers of furnishing fabrics and household textiles normally produce one main range per year.

In every segment of the market, the higher-volume manufacturers work further ahead of the retail season than the companies producing lower volumes (at the higher end of the market). Generally, smaller operations do not require the long lead-times that are necessary for large-volume production runs.

2.3   Printed and constructed textiles

Textile designers may be categorised by the types of product or fabrics for which they design. Printed textiles are often considered to include fabrics patterned by dyeing techniques as well as those where the design is applied to the fabric by a printing process. Constructed textiles include woven textiles, knitted textiles, lace and carpets.

2.3.1   From sketchbook to fabric samples
The design process is about the realisation of ideas — ideas that are transformed into tangible products. The designer will start with some sort of brief describing the project. For the in-house designer, the briefing may be very informal as their experience within

the company will usually mean that they will already have an understanding of the requirements of the range. They may, indeed, be the person setting the brief! Hopefully, any brief, formally or informally delivered, will have all the information needed, but if not, the designer must be ready to ask questions. Who is the customer? What is the price range? What are the manufacturing possibilities in terms of machine availability? The designer needs to get as much information as possible.

Designs come from a variety of different sources, and different designers will have different methods for developing design work from initial ideas. Most designers keep some form of sketchbook — something they carry with them and use to record and store ideas to call upon as a source of inspiration in the future. This may contain drawings and photographs, magazine cuttings and fabric pieces. Often, a brainstorming session will be used to give inspiration and generate a wide variety of ideas. Some of the considered themes and ideas will be selected and subsequent artwork based around these. While the way these ideas are used will be individual to the designer, much of the process is similar for many different types of designers.

2.3.1.1 Yarn design

Yarn designers have to know what machinery they have available to them and what that machinery is capable of producing. They will usually start with ideas collected on their travels; colour ideas, texture ideas, looks and moods that they see as becoming important in the future. They should have an understanding of their existing and potential customers and what they want.

The development of a range of yarns may involve the creation of a completely new yarn using a newly-developed fibre. It may involve the tweaking and fine-tuning of existing production ranges to meet the requirements of a future market. An existing yarn quality may require to be re-coloured so that it can be used to service a fashion market. Some ranges of yarns will run from season to season servicing a classic market area. For example, a range of knitwear yarn for a manufacturer of schoolwear will include the classic school colours — navy, maroon, grey, bottle green, bright red and bright blue.

The knowledge base that the yarn designer will have built-up is the springboard from which they work, in conjunction with technicians and development personnel, looking at the overall range, determining qualities and the types of yarns appropriate to their market area. The selling cost of the yarn must also be considered.

Colour is very important and yarn designers will often subscribe to colour prediction publications to help them determine their colour palettes. The same colours or shades of colours will often run through the different yarn types. Co-ordination is as important in yarn design as in any other area, helping the customer to see how the product can be used and often prompting more sales. Yarns may be designed to work with other qualities in the range.

Sample yarns will be specified and made using different components and in different colourways until the designer is happy with the handle and appearance. The yarns may have to perform in specific situations and they will have to be tested to ensure that any such performance requirements are met.

2.3.1.2 Weave design

After the briefing meeting where the work that is required is established, and after initial ideas have been collected, one of the first things that a weave designer will do is to establish a colour palette. This colour palette may change as design work develops but such changes are normally minimal. A colour palette will usually consist of groups of colours, chosen with regard to trends and predicted directions. The way these colours work together will be important. A degree of flexibility is usually desirable as economic constraints will normally demand that stocks of yarns (both for sampling and production) are kept to a minimum.

Once an initial colour palette has been decided upon, suitable yarns are then selected. These will normally be selected from ranges offered by yarn producers and frequently from the colours in which the yarn producers have decided to run their ranges; sometimes, however, colours will have to be specially dyed.

From their initial paperwork, weave designers will usually take their design ideas and develop them on the loom. Sample warps will be made up and different weave, colour and yarn combinations tried. Shape and pattern are important but these might well evolve more with the weaves used. The way woven designs repeat is determined by the way the yarns are threaded up and the way these threads are lifted. This is a very integral part of the weave design process, and a good understanding of fabric structures is vital to allow the weave designer to be as creative as possible within the constraints set by the type of loom that is being used and its patterning capabilities. Sample blankets, or 'section' blankets as they are also known, consist of different sections across a warp. These sections may vary in colour, yarn type and/or weave. Initial sample warps will often have several warp sections with different drafts, and different warp patterns and yarns. Different lifting plans (which, along with the way the loom is threaded, determine the weave) will be tried out as will a variety of weft patterns and yarns, again in sections. As well as true designs where the planned warp will be woven with the intended weft, there will be many areas where crossings occur. These are sections where warp and weft patterns combine more by chance than by design. Areas of interest will be developed and hopefully an initial sample blanket will provide plenty of ideas for subsequent development. Having considered initial ideas, further sample blankets will be developed; these will show full repeats of the proposed fabric designs and will often have sections in alternative colourways.

Some weave designers will work on computer-aided design (CAD) systems, using these to try out ideas before the weaving stage. Many of the sophisticated weave CAD systems currently available, as well as showing representations of what the fabric will look like, also have facilities to show proposed weave designs in situ, on garments and on other products. However, in practice it is still usual to weave actual fabric samples of those ideas which best meet the brief. The way fabric feels, its handle, is still a very important factor in fabric selection.

With a selection of ideas chosen, the making particulars or fabric specifications will be recorded so that these designs may be produced again. The yarns and colours used, the draft (the way the loom is threaded), weave, lifting plan (the way the warp threads are raised), warping and picking plans (colour and yarn arrangements in the warp and weft directions) all need to be recorded, as do any subsequent process such as applying a fire-retardant finish or raising.

Throughout the weave design process, the designer is constantly having to make decisions, on colour, yarn, weave, size of repeat, etc. These decisions are made from a knowledge base that includes intended selling price, customer preferences, the range as a whole, co-ordination requirements, performance requirements, and so on.

Once the design ideas are developed through into fabrics, these will usually then be presented in-house to the sales team, and then to customers.

2.3.1.3   Knit fabric and knitwear design
Again, as in weave design, the design of a new season's range will usually start with the development of a colour palette. This will come from initial ideas generated in response to information from a briefing meeting. Suitable yarns will then be selected. The working through of design ideas is very much restricted by the type of knitting machine being used and the type of structures that can be created on it. Inspiration from drawn work will influence the textures, colours and proportions of a knitted fabric. Creating visual representations of sketchbook work is only possible when machinery with jacquard patterning capabilities is used.

Some designers will put their initial design ideas straight into graph form, ready for translation by a knitting mechanic. Others may do more paperwork, perhaps developing fabric and garment styling ideas side by side.

The fabric ideas worked out, the designer will then usually decide which yarns should be tried in which fabrics. It is usual to knit several swatches to enable the designer to see how any fabric looks in different yarn and colour combinations.

For the knitwear designer, concerned with the design of final garment as well as the fabrics, styling ideas must then be worked out. The designer of cut and sewn knitwear will probably sketch out styling ideas, and the next stage will be to cut a pattern. A first sample will be cut out and made up, pressed, measured and tried for fit. Any necessary alterations will be made and a second sample made to check that these alterations and the final specifications are correct.

To help knitted fabric and knitwear designers, there are specific CAD systems available which allow representations of fabrics to be viewed before knitting. They also often allow fabric ideas to be mapped onto garments, and these facilities should cut down on actual sampling time. In practice, this is not always the case as the ease of changing things on the computer often results in requests for more initial work giving more choice.

Once the garments and colourways that will form the range are decided upon, the range is then ready for presentation to sales teams and customers.

2.3.1.4   Print design
The print design process can take many different forms. With modern technology, sketchbook work can be taken and re-created on fabric very much as it appeared originally.

Print designers, after initial research, will usually work their ideas through on paper. Any medium can be used in the artwork for print design, but materials with which designers can achieve facsimiles of particular desired effects in manufactured fabrics allow design problems to be more accurately resolved in the studio before the design is put into production.

Print designs are sometimes worked in repeat from the start but often they are designed in balance and put in repeat later. A balanced design is called a 'croquis' (a French word meaning *sketch*). A croquis should give the impression that would be seen if a frame were placed over any section of the finished cloth. Although not in repeat, a croquis will give the feeling of being in repeat. The most common repeat structures are simple block repeats, half drops and bricks (see Chapter 4).

Print designs for apparel fabrics are usually designed in croquis form with the repeat sizes varying enormously. Upholstery and curtain fabrics have standard repeat sizes so these are often designed as a repeat pattern with the size being carefully considered from the beginning.

Often, the involvement of a print designer will end with the production of the design on paper, the preparation of colour separations and the approval of these usually being undertaken by the printer. However, in-house print designers will usually be responsible for approving the strike-offs (sample prints on fabric).

Again many print designers (particularly those working for large multinational companies and corporations) will often find themselves working with CAD systems. Such print systems allow designers to develop and expand design ideas readily. Design ideas can be scanned in, and repeat, cut-and-paste and other functions allow ideas to be manipulated quickly. Many systems also have colour separation facilities to aid in the preparation of screens. Some systems even allow sample fabrics to be produced, and this is a very valuable tool, making it much easier for print designers to take their ideas right through to fabric.

2.3.2   Design adaptation and modification

As well as producing ranges, textile designers will have to adapt and modify existing designs, design ideas and sketches submitted by customers. Here, textile designers are using their creative and technical skills to translate ideas into reality. They must, of course, be careful not to copy or adapt designs that have copyright, or are registered, without the express permission of the copyright holder or design originator.

2.3.3   Examples of textile design briefs/problems

An in-house designer working for a manufacturer of printed fabrics may be asked by a customer selling babywear to develop a small design with chickens as the motif. The babywear company may also employ a designer and they might have put together rough sketches and colour ideas to give an idea of what is wanted. In this case the brief will be fairly specific.

A company may have seen a range of designs for duvet covers that they like. However, they may be unhappy with the colourways being offered. An in-house designer working for the printer may be asked to re-colour the designs, producing colourways that are considered more suitable. Perhaps these will have to co-ordinate with a range of lampshades being developed in a specific range of plain shades.

A knitwear manufacturer might find themselves approached by a company which has a sweater that they are manufacturing overseas and are currently importing. The company is, however, experiencing real problems with quality and late deliveries. Essentially the company wants the knitwear manufacturer to re-create their sample sweater to a specific price point. The designer's job in this case would be to source suitable yarns and develop specifications for the production of a garment that is essentially the same as the original sweater in terms of quality and appearance but which also meets the pricing requirements.

A company might have a fabric that is currently selling very well — their 'best-seller'. The marketing director and his team require a 'new' version of this fabric, perhaps more up-to-date and with a co-ordinate. They are happy with the existing colourways but are open to suggestions regarding colour and scale. As their market area is becoming increasingly competitive they would also like to see the new version having elements of 'added value' if at all possible. This might mean repositioning the product in the marketplace or developing a new product that acts as a co-ordinate with the current best seller. In this case the textile designer's job is to develop a fabric using the existing colourway that fulfils the company requirements.

2.4   Summary

Textile designers have to design and produce, to an agreed timetable, an agreed number of commercially viable fabric designs. There is a huge diversity of designers involved in textiles; from colourists to knitwear designers, yarn designers to print designers. Stylists put together ranges while repeat artists put designs into suitable repeat layouts. Textiles are usually designed well ahead of the season that sees the products containing them in the shops. For apparel there are normally two ranges produced per year (Autumn/Winter and Spring/Summer) while furnishing fabric manufacturers normally produce one main range every year. Textile design can be categorised as constructed and printed. Different textile manufacturing processes will require the designers to be involved in different ways of designing, although much of the textile design process is similar. As well as designing from their own ideas, textile designers will often be asked to adapt and develop existing designs and ideas from their customers. There are designers who work in-house as staff designers for specific companies, while some designers work on a freelance basis and others work as consultants, advising companies on their design policies. There are advantages and disadvantages in all these systems, both for the designers themselves and for the companies employing them.

Bibliography

McIntyre, J.E. and Daniels, P.N. (eds), *Textile Terms and Definitions*, 10th ed., Manchester, Textile Institute, 1995.

Taylor, M.A., *Technology of Textile Properties: An Introduction*, 3rd ed., London, Forbes Publications, 1990.

Yates, M., *Textiles: A Handbook for Designers*, rev. ed., New York; London, W.W. Norton, 1996.

3

The textile design function

3.1 The activities of a textile designer

The overall purpose of a textile designer is to design and produce to an agreed timetable, an agreed number of commercially viable textile designs.

Research into what designers do has identified several different activities. The activities listed below relate to an in-house fabric designer engaged on branded, contract and general work.

- Deciding what to design.
- Producing original design ideas.
- Developing design ideas through to a form suitable for initial sampling.
- Supervising the production of original fabric samples and keeping suitable records.
- Submitting samples to customers, to company or to company selection systems.
- Adapting and modifying designs from sketches or fabrics submitted by customers (or obtained from within the firm or group) to meet a price or other restriction.
- Obtaining the acceptance of a certain proportion of the original sample fabrics by the customer.
- Controlling the production of sample ranges in consultation with the customer's technicians, technologists and buyers.
- Establishing production specifications.
- Ensuring the production of sample ranges by certain agreed dates.
- Reporting to the company on contacts with customers, competitors, exhibitions, developments by fibre and yarn producers, etc., in order to augment market intelligence.
- Controlling the systems and procedures for the storage and retrieval of designs and samples within the company.
- Working within an agreed departmental budget.
- Controlling the supply of materials and equipment for design and sampling.

All textile designers will certainly be involved in some of these and should have an understanding of all of them.

3.2 How design work is done

Design ideas are developed through in different ways, depending upon how an individual designer likes to work, upon the product being designed and upon the manufacturing process/es. From initial paperwork, designers of constructed textiles may take their design ideas and develop them through on the loom or knitting machine. Designers of printed textiles will usually develop their ideas right through on paper.

'Much design work is carried out in a very direct and informal way. The degree of formality becomes a function of scale and the number of interests represented.' [1]

- Design is an investigative process; it involves research. The first stage in any design exercise is normally an enquiry into what the client (or potential client) requires; their needs and expectations.
- Design is a creative process; it involves art and aesthetics. Designs can be copied or invented. A design problem is solved with the help of know-how, ingenuity, pattern recognition abilities, lateral thinking, brainstorming, etc.
- Design is a rational process; it involves logical reasoning in the checking and testing of proposed solutions, information analysis, experimentation, field trials, etc.
- Design is a decision-making process; it involves making value judgements.

The selection of particular combinations and configurations, layouts or shapes involves considerable uncertainties. These are resolved by estimating the values that are likely to be placed on the various major alternatives. In this respect the task of the designer is closer to that of a manager rather than a researcher, scientist or artist. Estimating the social consequences of various design alternatives also falls into this category.

Solving design problems involves a mixture of the intuitive (having ideas) and the systematic (rational scientific appraisal). There is no single solution to a design problem, rather there are many solutions although some may be better than others.

3.2.1   The design process

The design process usually starts with a requirement or desire for a new item or product. Research will usually be carried out then to find out as much as possible about this need and about the role or function the new item or product is to have. Ideas generation is the next stage when various alternative initial ideas are conceived. These initial ideas are then usually developed through until the designer is happy to offer them as proposals to meet the initial need. In the early stages, alternative ideas will often also be presented. These proposals will be considered and perhaps modified. A decision is then taken as to the best solution to the design problem and the necessary specifications and instructions will then be given. (See Fig. 3.1.)

3.2.2   Planning design work

Any design project will usually need to be completed by a certain date, and more often than not designers will find themselves working on more than one project at any one time. Efficient project management and efficient time management are obviously both desirable.

A project entails one or a number of individuals working together over a period of time to achieve an agreed goal or outcome. Any project needs to be completed on time, to a standard and within a budget. A design project, as any other project, needs to be planned and managed efficiently.

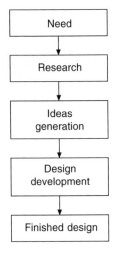

Fig. 3.1   The design process.

3.2.3   The planning process

Planning can be described as managing and controlling events to achieve a goal and making the best use of resources.

Planning is achieved by making decisions about what is to be achieved (aims) and how best to get there (who does what, when and how). While planning takes a little time, lack of planning can result in the waste of a lot of time, the expenditure of much money and the generation of considerable stress and tension.

A plan gives what could be described as a set of route markers that allow checks to be made as work is undertaken to achieve the goals. Regular checking of current position against plan identifies any deviations from course which can then be remedied. The more detailed the plan, the more route markers there will be. A plan helps in the controlling of events rather than events determining direction. Very few jobs go exactly according to plan, but if there is no plan there is no way to keep control and track progress.

3.2.4   Objectives

An objective describes something to be achieved in the future. The essence of an objective is that it defines what is wanted without describing how this is going to be achieved.

Without objectives people do not know where to invest their time or resources, and they can drift in an unco-ordinated way. When objectives are unclear, it is impossible to make effective plans and people will not be committed to action. Despite the obvious importance of objectives many people find it difficult to set them.

3.2.5   Identifying the aims and objectives of a design project

The aims and objectives of any design project are usually identified at a briefing meeting. This is when the designer finds out what is required of them. Many clients often have little idea of what they actually do want so it is up to the designer to get as much information as possible; this is usually done by asking questions.

3.2.6   Checklists

Using a checklist can be very helpful at many stages in the design process. For example, a checklist for a briefing meeting would simply be a list of questions, prepared beforehand, that the designer will use as a prompt (see Chapter 6). Enough knowledge of a situation is required to enable a checklist to be developed that covers the right questions. Use of a properly prepared checklist should ensure that the designer does not forget to ask for any necessary information and this should help avoid misunderstandings that might create problems later in the design process.

3.2.7   Project planning methods

Different methods can be used to help plan design projects. The simplest is to arrange elements of the plan in a logical sequence. Working from the brief, designers determine what is required to fulfil this and plan a logical sequence of events to take them through the design process to a design solution that satisfies the brief.

3.2.7.1   Backwards planning

Often, the best way to ensure that design work achieves the requirements of the brief is to start by examining the expected final products or outcomes. These products or outcomes need to be listed and agreed. Having considered what it is that is to be achieved, the processes or work to achieve the outcomes can be determined. Finally, it must be decided what inputs are necessary to enable the work to be carried out successfully. (See Fig. 3.2.)

Fig. 3.2   Backwards planning.

Outcomes: The planned result of the project, equal to the aims and objectives of the project, to achieve the quality or standard required.

Processes: The work needed to be done to achieve the output deliverables or outcomes.

Inputs: The resources in terms of individuals and their skills, time, materials, equipment, techniques, etc., necessary to achieve the outcomes.

For example, a designer is asked to design a range of printed curtaining fabrics. The *outcomes* would be a specified number of fabric designs, with colourways, for a specific time, to sell at a specific price point. The *processes* would be initial research of the market (to see what is currently selling and what the competition are doing), the formulation of initial design ideas, the development of these ideas, the selection and production of the specified number of fabric designs (with recommendations for print base fabrics) to be included in the range, colourway development and selection, and the specification of final fabrics for production. The *inputs* would be the designer with the required skills and knowhow, the time they would require and the materials and equipment necessary.

3.2.7.2   Gantt charts

A slightly more sophisticated way of planning is to use a Gantt chart. During a design project, several stages can be under way at the same time. A Gantt chart is a simple horizontal bar chart that graphically displays the time relationships of the stages in a project. Each step is represented by a line or block placed on the chart in the time period in which it is to be undertaken. When completed, a Gantt chart shows the flow of activities in a sequence, as well as those that can be under way at the same time. (See Appendix A.)

Gantt charts can also be used to chart actual progress, by drawing lines in different colours to show the start and end dates of each step. This allows easy assessment of whether or not a project is on schedule.

3.2.7.3 Network analysis

This is a generic term used for several project planning methods, of which the best known are PERT (Programme Evaluation and Review Technique) and CPA (Critical Path Analysis). These are more sophisticated forms of planning than Gantt charts and are appropriate for projects with many interactive steps.

3.2.7.4 Project management

Managing a project involves co-ordinating activities so that they run according to plan. The progress of a project should be monitored and measured against the plan. When deviations occur, corrective action should be taken.

3.2.8 Time management

All projects are time-bounded and for any practising designer it is desirable (if not essential!) to make the best use of the time available. Time management is simply making the best use of time to achieve what is necessary. To effectively manage time, goals and time limits need to be set. What is required? What has to be achieved and when? How efficiently do these goals have to be met? It is only against set targets that success can be measured.

The way any individual uses time is unique to that individual. Some people use time as chunks into which they can fit certain activities, all neatly stacked. Others have no clear view of time, selecting activities at random or changing priorities to suit the current crisis. There are strengths and weaknesses in both ways and it is of value to consider both because many people will alternate between the two, depending on the jobs in hand. The way designers use their time will be different, since everyone has their own pace of work and their own rhythms, with different peaks for different activities. Where possible, work should be done at times to suit an individual's own speed and their own way of organising and completing activities.

Managing time costs time. To sit down and plan the best use of time is an investment. It takes time to learn to use software packages for word processing and database management but their use is a huge investment for future time management.

The time spent on any activities should be considered afterwards and evaluated. Can any lessons be learnt to plan more effective use of time in the future?

The Pareto principle: 20% of what you do yields 80% of the results.

Targets that will result in a high pay-off should be identified. Constructive avoidance is when time is spent on work that is neither important nor urgent in preference to urgent and important work.

Time management involves:

- *planning tasks –*
  getting information, assembling relevant facts, skills, experience, resources, establishing what is known or what needs to be known, processing the information, considering options and identifying the risks involved,

- *stating what has to be done* –
  scheduling what will be done, how, where, when and by whom.
- *getting things done* –
  doing things and monitoring progress against checkpoints/standards,
  reviewing outcomes by assessing the results achieved in relation to the aims, determining if more needs to be done, analysing successes and difficulties so as to plan
  for improvement.

3.3   Range planning

A range is a group of products offered for sale. These products will be the designer's
answer to a brief. Having the right products in a range is very important in terms of
how well that range will sell. (See Fig. 3.3.)

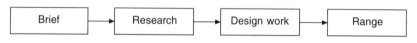

Fig. 3.3   Range development.

Good design adds value to a product. With products of equal quality and price, the
design will be what differentiates. What does the designer need to know or make decisions
about when planning a range? Or, put at its most simplistic, how does a designer decide
what to design?

All designers are designing for the future and are influenced by trends in their product
area. The whole area of forecasting (see Chapter 5) is one where much money is invested
to 'get it right' and much money can be lost if a designer gets it wrong.

As well as needing to understand forecasting, a designer needs to know other information when planning a range. This includes:

- type of products,
- customer/market areas,
- price points,
- manufacturing capabilities,
- quality required,
- number of designs,
- number of colourways.

It is usually the job of the design manager (or stylist) in a studio to plan the ranges.
There will have been some discussion with sales, marketing and production about all
or some of the factors listed above—essentially to build up a full picture of what criteria
the range must answer.

The type of designs depends on the products being designed for, the customer profile,
price points for sales, manufacturing plant availability and quality. The number of designs
and colourways in a range is controlled by several commercial factors including the
amount of capital the company wishes to tie up in the range, the raw materials that are
needed, the amount of stock that can be carried, and the production capacity, i.e. the
number of machines available for printing/knitting/weaving the range (some machines
may be tied up with specific customer orders rather than general range production).

3.3.1   Research

After the briefing meeting, the next stage in the design process is usually research: finding information about products and markets, production processes and techniques, etc.

Information can be collected in different ways:

- By waiting to see if information presents itself (gathered by chance from contacts and perhaps by reading).
  This is unsatisfactory if there is any form of time limit on when the answer is required.
- By asking questions.
  This is appropriate if it is known who to ask and what to ask. Known as primary research or field research, this is where designers find information directly for themselves; by going out and looking at the shops and visiting trade fairs and exhibitions, by asking relevant individuals and groups, by using mailed questionnaires, by setting up focus groups and by interviewing people.
- By looking up what is required.
  This is appropriate if it is known where to look, how to look it up and what exactly is being looked for. Known as secondary or desk research, this is where information is gleaned from work already carried out by other individuals or groups. Such research might include looking at market research information obtained through government censuses and large market research programmes.

3.3.1.1   Information search

To ensure that the required information is found and time is not wasted, the collection of information should be systematic. The aim should be to find key information. Identify what is not known. As far as possible, vague requests for 'information on' some problem area should be broken down into questions defining what it is that is not known and what needs to be found out. The questions that need to be answered should be identified and listed. The level of answer required should be established (there is a balance between the importance of what is being looked for and the time and effort spent looking) and a checklist of likely sources of reliable answers to questions can be prepared. It is sensible to keep control over any search for information by setting limits and deadlines and by using expert opinions and expert literature searchers to identify the most promising sources. It is important to stop as soon as there is enough information to move forward and it is essential to keep accurate references.

3.3.1.2   Getting information from other organisations

There are many occasions when designers need to find out information from other organisations. Sometimes, the information sought may have to be bought, sometimes it may be free — part of a marketing strategy on behalf of the supplier. However, whether visiting a new supplier to source yarn or buttons, or telephoning to check prices of fabrics, it is important that designers know exactly what they want so as not to waste time. Before any visit, the information required should be clearly identified. It is essential for any designer to be prepared, and establishing a checklist of questions that need to be answered is a good way of making sure that all relevant information is obtained.

When seeking information about market areas and products, it would be usual to employ professional market researchers. However, sometimes designers may prefer to carry out some market research themselves. In either case, well-constructed questionnaires can be very useful, and when market research is being carried out by professionals the designer's input should not be overlooked.

3.3.1.3   Questionnaires

Questionnaires should be designed in such a way that they take a minimum amount of time to fill in. A busy company director or buyer is much more likely to complete and return a questionnaire if it is clear to read, to the point and does not take up too much of their valuable time. Formulating questions in such a way that they can be answered by ticking boxes is much more likely to elicit a positive response than having questions that require written replies.

3.3.2   Ideas generation

With some initial research undertaken, the designers need to be able to generate suitable ideas to help them answer the brief. Design ideas can come from almost any source, but sometimes designers may find difficulty in thinking of suitable sources and themes for work.

3.3.3   Brainstorming

Brainstorming is a very useful method of generating ideas and can be used at many stages in the design process. In its most formal sense, it is a group participation technique for generating a wide range of ideas in order to tackle a stated problem. In a less formalised way, an individual or pair can use a brainstorming session to generate ideas. What follows is a description of brainstorming at its most formalised; such a rigid approach, however, may well inhibit ideas.

1) A problem statement is formulated. Too vague or too restrictive a statement should be avoided.
2) The group of people to participate in the session is selected. The group should ideally include some people familiar with the problem area. Small groups of about 4–8 people are best.
3) Five or ten minutes are allowed for group members to write down their first ideas in reply to the problem statement.
4) Group members are encouraged to continue writing down new ideas whilst each person in turn reads out one idea from their set.
    The session rules are:
    (a) No criticism is allowed of any idea.
    (b) Crazy ideas are welcome.
    (c) The more ideas the better.
    (d) Ideas should be combined and built upon.
5) After the session, the ideas are evaluated.

3.4   Range development

So how does the design manager go about putting a new range together? Once a basic colour palette has been decided, this will help the ideas being developed to work together as a total package. Colour co-ordination does not mean that every design has to be in exactly the same colourway, but colours should be related.

New fashion colours should be looked at alongside any existing palette to make sure that these work as part of the range. Decisions have to be made as to whether best-selling designs from a previous range are going to be carried on. If a design was well received and is still selling well, it makes commercial sense to carry it on. It is hard to argue with sales figures. When continuing a design through to the next season, it must be decided whether to change the colourways to bring it in line with the new season's colours or whether to continue the design in the existing colourways and add some new colourways. Re-colouring can add to the sales of a proven design.

Areas and themes will have been identified according to the customer requirements, as will the appropriate number of products to fall within them. Having established the number of products and colourways for each, initial ideas for the range can be put together. At this development stage it is very necessary to maintain an overview of the whole range. If it is a range of apparel fabrics that is being developed, then there might be part of the range aimed at menswear and part at womenswear. An important existing customer may be a pyjama manufacturer and so designs suitable for that customer must be included.

Designs may be developed in-house or bought in, or a mixture of both. Designs may come from archives of designs that have already been developed and sampled. A few designs can be assembled around each theme or group and any areas where it is felt designs are missing will prompt more sampling. Production capabilities and machine loading must be considered. For example, there is no point when designing a range of knitted fabrics to produce all the fabrics for this on one specific machine type if the manufacturing company owns several different types of machines, producing quite different fabrics. It is only when all such considerations have been taken into account that the final selection of the designs to be included in the range will be made.

Once fabric designs are chosen, suitable colourways will be selected, and if necessary, sampled. Colourways may be balanced or not. Balanced colourways are when the colours change but the relationships of the colours within the design stay the same. Unbalanced colourways are when there are no similar colour relationships between colourways. (For example, a three-colour print design has a dark blue ground, large flowers in mid-blue and small flowers in light blue. A balanced colourway of this would be one that had a dark green ground, large flowers in mid-green and small flowers in light green. An unbalanced colourway would be one that had a bright yellow ground, pink large flowers and dark blue small flowers.)

For a range to work as a total package, themes and concepts can also be used as well as colour and colour co-ordination. For example, Dorma, a UK bedding manufacturer, produced a very successful range of bedlinen based on the illustrations in the book *The Country Diary of an Edwardian Lady*. This range was so successful for Dorma that several seasons later a further bedding range was developed from the same book, but this time using a different colour story.

Themes can be used as a base for designs and colour co-ordination. Themes and concepts can also help with publicity, being taken through to brochures and in-store presentations.

Many fabric ranges will include plain colours. In most product ranges there are basics that have to continue, and decisions have to be made as to which colours will be carried on and which will be dropped. Often, these plain colours are related to other products in the same area; for example, towels relate to bathroom fittings, telephones to paints and wall coverings. There should be some co-ordination between plain and patterned ranges. This allows customers to buy from both and thereby increases sales.

Range planning and development is very important. The brand range is what a company sells itself on. The image created with the range will be the image of the company for the next twelve months or so until the next range.

3.4.1 Design development

The way design ideas are developed will depend on two main factors; how an individual designer prefers to work and what is being designed for. Designers will usually try out different ideas, taking the best of these through into fabric or, in the case of print design, through to designs painted-out to size and in repeat. Computer-aided design (CAD) systems can be very helpful at the sampling stage, allowing different ideas to be tried and fabrics to be simulated without the expense of fabric sampling.

3.4.1.1 Sampling

Fabric sampling is an expensive process. Very often, machinery has to be taken out of production for sampling, and time that is not being used to make saleable products on expensive machinery is costly. Money tied up in sample stocks is also money that cannot be recouped.

3.5 Range presentation

The introduction of new products to a company's product range used to be done in a very arbitrary way, with little planning or co-ordination. In the early 1970s however, things began to change. Many companies closed due to world-wide recession. Those that survived had to work harder to maintain a place in the market. Ranges had to be well presented, with marketing becoming increasingly important. For example, for yarn manufacturers and spinners it was no longer sufficient to show new yarns just on cones. Ranges had to be carefully planned, the presentation of the range to the customer became very important and the presentations began to include how a range might be used. Spinners started to employ designers to put together trend ideas. Different yarn qualities within a range would often be colour co-ordinated to help the customer maximise use, and fabrics that illustrated the current trends would be made up from the yarns in the range.

Presentations of a range are often both written and oral. There will normally also be drawings and photographs illustrating how fabrics from the range work together, how fabrics work when made up into or incorporated into products and how these products work with other, related products. Presentations often reflect the theme used for the design work.

For a presentation to be successful, the designer must have answered the brief. If the brief is not clear, it is difficult to solve the design problem. At a briefing meeting, the designer is trying to get as much information as possible. Everyone who briefs a designer will have a preconceived idea of how a job will turn out and the designer's job is to find out what this is. It is therefore important that presentations should be made to the person who gave the briefing.

Presentation of the final range to the person who asked for the work to be undertaken is, however, not the only type of presentation a designer will be called upon to contribute to or to make. Once the range has the approval of the relevant directors or company board members, the designer may well be asked to make presentations to sales teams, agents and customers. Designers are called upon all the time to give presentations of their work and so an ability to present themselves and their work well is very important.

3.5.1 Presentation of initial design ideas

At various stages of the design process, designers are often called upon to present their ideas to their colleagues, managers and directors. Designers have to look the part, they must present a 'designerly' air and should show themselves as being organised and capable. The best presentations are simple and to the point; this is as true of visual presentations as of written and oral presentations.

3.5.2   Presentation of design and artwork

Designers are visual people. Their work will often need little verbal explanation, clearly communicating the intention by virtue of visual impact. The way such work is presented is very important because good presentation can show a design solution to advantage while poor presentation can hide and distract from good design. Presentation is itself an exercise in design.

Good presentation should be suitable and professional. It should be suitable in that the presentation style and techniques should show whatever is being shown (be it fabrics, yarns, artwork or drawings) to its best advantage; it should look as good as it possibly can. All presentations should be professional in that they should look neat and tidy, deliberate, considered, planned, well designed, appropriate, consistent, related and reflect the work presented. Poor presentation is chaotic, messy, dirty, disorganised and inappropriate. (See Appendix B for tips on presenting work.)

3.5.3   Visual presentations made by textile designers

The types of visual presentation of their work that textile designers find themselves making are for a variety of purposes:

3.5.3.1   Mood/theme boards

These boards are collections of ideas around a mood or theme. The ideas may come from magazines, postcards, colour swatches, yarn or fabric pieces. Presentation of mood boards should be simple and suitable; the style of presentation should reflect the mood. A sporty mood board will be presented differently from a mood board reflecting luxury and wealth.

3.5.3.2   Initial design ideas

It is often appropriate that designers themselves present their initial design ideas to a customer or client. This will allow the client to feel that they have some input and also should prevent unsuitable work being done. The client may be in-house or external. An external presentation may be more formal than an in-house presentation, but the presentation should in both cases be appropriate and show the ideas to their best advantage.

3.5.3.3   Finished design work as the solution to a design problem

Finished design work is presented to show the design solution to the brief. It will often show the development of ideas through to the final design work.

3.5.3.4   Work to show capabilities and potential

Designers will build up a portfolio of art/design work to show what they are capable of. Art/design work should show a variety, a selection of the best work showing how ideas are worked through. It should be neat and tidy, and presented in a coherent manner. It should be easy to look at. When making decisions about what should be included in a portfolio, it is necessary to consider who will be looking at the portfolio — how much time will they have? What will they expect to see? etc.

3.5.4   Publicity/promotion/packaging

If a range is to sell well, it must be presented in a simple and understandable way to the non-designer. Publicity can contain photographs of furnishing products in room settings, emphasising the importance of colour co-ordination and pattern co-ordination. It can be seen how the range is related to other products in the same room area. Publicity for fashion garments will show how the garments in the range can be used with other garments.

If the range has a name or theme this can be emphasised, for example through publicity with a distinctive logo. A logo can be used on packaging and on any labels. If the product is not fully visible when packaged, then a photograph will be needed. This technique is often used for duvet packaging; the folded duvet will be packaged with a photograph of the duvet *in situ*, allowing an appreciation of the full design. Such photographs may well also be used for in-store displays. If the product is targeted as a gift then the packaging should be especially attractive. Soap may be sold with a towel. Products sold in a basket mean that the packaging becomes a product in its own right.

3.5.5   Store display

The days have gone when, in a department store, the bathroom fittings were in the basement and towels on the fifth floor, with no hint of a relationship. Concept selling offers many related products in an area, and bedding, curtains, lampshades and other items for the bedroom may be displayed together, co-ordinated by colour and a common or linked theme. Such selling techniques make it easier for the designer to get across the message of co-ordination and for the customers to see how the product will work in their home. However, presenting merchandise in this way is not always easy for department stores, where buying systems have been traditionally broken down into product areas.

3.5.6   Presentation to customers

This is when the designer shows the finished range of products. The presentation may be to selectors from a major high street store group, or it may be to customers from within the designer's own company or group of companies. The designs may be either the work of the designer doing the presentation or the result of several designers and others, such as technicians and sales personnel, who have all had an input. Design is very much a team activity with the designer co-ordinating the design development, whether this is for one fabric or a range of fabrics.

Designs have to fit into a customer's market area so designers need to be familiar with their customer's current ranges. They have to try to identify gaps in product and colour ranges and their ultimate design work should be to fill these gaps.

Designers need to try to identify new directions for their customers. It is important that new ranges lead the way rather than follow so that customers come back for new ideas. Storyboards are frequently used to illustrate points in such presentations. Feedback from the customer on designs and directions should be encouraged to ensure that the customer is involved and has a degree of ownership, and that the brief is answered. This also helps establish good working relationships.

3.6 Summary

The textile design process can involve a designer in many different activities, from deciding what to design, through controlling the production of sample ranges and establishing production specifications to controlling design storage systems. The design process is investigative, creative and rational, and involves decisions being made. The design process starts with a need, involves research, ideas generation and design development, and ends with a new design. Design projects have to be completed to schedules and within set budgets, so efficient project management and time management are important. Planning is about managing and controlling events to achieve a goal or goals. The aims and objectives of any project require to be identified. This is usually done at the briefing meeting, and here checklists can be a helpful tool. There are different methods that can be used to help in planning projects. These include backwards planning, Gantt charts and network analysis. How an individual uses time is unique to that individual. Planning how to best use time is crucial. Research is important to the designer — both primary and secondary research. Information can be gathered from a variety of sources and in different ways. Generation of ideas can be aided by brainstorming and this can be carried out formally or very informally. Range planning, range development and presentation are vital functions of the textile design process.

References

1. Potter, N., *What is a Designer? Things, Places, Messages* 3rd rev. ed., London, Hyphen, 1989.

Bibliography

Adair, J., *Effective Time Management: How to Save Time and Spend it Wisely*, London, Pan, 1988.

Bond, W.T.F., *Design Project Planning: A Practical Guide for Beginners*, London, Prentice-Hall, 1995.

Davidson, J., *Effective Time Management: A Practical Workbook*, New York; London, Human Sciences Press, 1978.

Open University, *Managing Design*, Milton Keynes, Open University, 1989.

Yates, M., *Textiles: A Handbook for Designers*, rev. ed., New York; London, W.W. Norton, 1996.

4

The principles and elements of textile design

Textiles are frequently made to be decorative and are used to embellish and decorate both people and objects. The designers responsible for such textiles have to balance many factors when answering a design brief. What is the fabric for? How must it perform? Who is the customer? What are the economic restraints? How is the fabric to be produced?, etc.

Research into how people select products shows that colour and appearance are two of the most significant factors, with handle, performance and price coming lower down in terms of importance. Textile designers therefore need to have a good understanding and sensitivity to colour and aesthetics.

The same design elements and principles apply to design as to fine art, and an understanding of and an ability to use these are as vital to the designer as to the fine artist.

'Design is about relating elements, whether they are similar or contrasting and visually arranging an interesting unity with them. Shapes, forms, colours and texture all combine to become a unified whole which is commonly called a design.' [1]

Art and the act of creating are highly personal activities. In the first instance it may seem impossible to objectively analyse works of art. In different cultures and at different times, styles of art and even the media used to express art have varied enormously. There are, however, certain basic design features that appear to be natural and common to all art.

4.1 Design principles and elements

Design features, the elements and principles of design, can be called the language of art and design. Pick up one of the many books on design and these elements and principles will be there, although they may be detailed slightly differently in each. This is, in part, due to the way that in any one work of art no one element or principle can exist in isolation.

Designing can be defined as relating and visually arranging components or elements to create effects. Space, line, shape, form, colour, value and texture are the design elements with which artists and designers work to create a design. The design principles of balance, movement, repetition, emphasis, contrast and unity are what artists and designers do with the design elements to make the art form or design.

4.1.1 Design elements

4.1.1.1 Line
A line can be a mark made by a pen or drawing instrument or it can be any continuous mark that causes the eye to follow along its path. The viewer's eye travels along the path of a line because a line is longer than it is wide. A line moves, and as it does so it indicates direction. A straight line leads the eye swiftly across the picture plane but the eye travels more slowly when following the path of a jagged line.

Lines appear in different ways. There are curved lines and straight lines. These can be long or short, thick or thin, ragged, sharp, light, dark, simple or complex. Lines can be broken and yet have direction. Lines can be textured and can be coloured. Lines can be made from repeating similar elements, as diverse as dots or people, in a lengthways direction. Lines can be used to create form, to give depth. Lines can be carefully controlled to create optical sensations and can be used to project feelings of sensitivity and strength.

When someone looks at lines they try to fit them into something related to their previous experiences; for example, scribbled spirals may be interpreted as seashells, and a few lines can easily suggest an apple to someone who is familiar with this fruit. In some ways this can restrict the artist's or designer's freedom to make viewers see what is intended, although it does also open up new possibilities. Skilful artists and designers play with this tendency to see familiar forms in everything by using only a minimum of visual clues to evoke a far more elaborate response. Care must be taken, however, that abstract designs and patterns do not unintentionally suggest unpleasant or ugly forms to the viewer.

4.1.1.2 Shape
When a line turns and meets up with its start point, a shape is created. A shape or figure is a positive thing and occupies positive space. The area surrounding a shape is called the background or ground. It is a negative thing and occupies negative space. Shapes can be clearly defined with hard edges but often they are not clearly defined, which means that it is more difficult to see where shape ends and background begins.

There are a great variety of shapes to be found in nature. Many artists have drawn creative inspiration from natural shapes. Through their imagination, artists have invented new ways to use shape and communicate ideas. Shapes can be solid or opaque, linear, textured, coloured and outlined. Shapes can be transparent, revealing other shapes behind them. Similar shapes need not be identical, yet they can have a common relationship, which visually ties them together. Some shapes will command more attention than others, depending on their size, colour, value, texture, detail or their location in relation to other shapes. Tall shapes are elevating, long flat shapes express calmness, downward-pointing shapes activate the sense of falling.

4.1.1.3 Form
Form is structure. In art and design, form is the illusion of three-dimensional volume or mass seen in two dimensions. Careful observation of the forms around reveals that, in nature and in man-made objects, many forms can be described as combinations of the basic geometric structures; spheres, cylinders, cones, cubes and pyramids. Paul Cézanne, the impressionist artist, in his landscapes and still lifes, explored how these basic geometric shapes combined to make more complex shapes.

There are several methods that can be used to give an illusion of three-dimensional form and some of these are detailed below. Line, shape, colour, value and texture can all be used to suggest form by the artist on a picture plane, and by the textile designer within fabric.

4.1.1.4   Space

Space flows in, around and between forms or shapes.

A flat surface has only two-dimensional space. This means that it has length and width but no depth. It is impossible to create actual depth or space on a flat surface but an illusion of space, distance or depth is possible. There are many methods used by artists and designers to create this sense of space, and to convince the observer that there is space and depth when, in fact, they are victims of a type of visual deception. Objects placed higher up can create the feeling of depth or distance. Overlapping shapes can also create the feeling of depth, as can converging lines.

If, in two-dimensional art and design, space exists purely as an idea or concept, it leaves the artist or designer free to compress or stretch it as they wish, to portray the particular feeling that is desired.

Colours have an effect on space. Colours that are warm and bright appear closer, while dull or cool colours recede into the distance. A flat surface that is covered only with pattern can eliminate any feeling of space.

4.1.1.5   Colour

Under normal light conditions, well over 10 million different colours can be seen. Throughout history, the study of colour and the development of colour theories were frequently undertaken by artists and designers, and many artists have spent their lives attempting to understand colour. Seurat wanted to apply a scientific system to the methods of the Impressionists, and his Pointillist theories, where dots of colour from a restricted colour palette were used to create the impression of a wide range of colour, clearly show his fascination with analysing colours.

Colours communicate; some colours are associated with cold (blues and greys) and some with warmth (reds and oranges). Colour can convey the time of day, weather conditions and temperature, and even the time of year. Colours can be designed to blend in with the environment or to stand out. Frank Lloyd Wright saw architecture as growing directly from the earth on which it stood and therefore used the colours of the surrounding areas in his buildings while Richard Rogers and Renzo Piano, the designers of Paris's Pompidou Centre, used colour, not to blend in, but to code the different pipe systems for heating and other services.

In art and design work, the colour can be used in a natural or abstract way. Maps make use of both abstract and natural colour; natural colours are used when illustrating mountains, deserts and seas while maps showing political boundaries will use abstract colour.

Colour is a crucial part of all branches of design and design-based industries. Advertising agencies know that graphic information in colour will have a more profound effect than that in black and white. Colour grabs the attention span because memory recall from colour is quite pronounced.

Colour is used to promote corporate identity. Blue is used by many banks to give the suggestion of reliability, while the appetite colours of red and yellow are used by fast food chains. Green is frequently used to denote environmental friendliness while more subtle greens convey a feeling of upmarket status. The *American Express* card was originally launched in yellow but was a flop; however, when its colour was changed to that of the American dollar bill it became a success. The green livery of *Harrods*, the world-famous upmarket store in London, epitomises good taste and sophistication.

Forecasting colour trends is itself an industry, and fashion colours change with the seasons. Underlying these cyclical trends, however, are some basic colour preferences. In colour popularity tests, blue is frequently placed first, with red second. Blue ties are most popular, with red ties in second place, and blue and red cars regularly occupy top positions, only being outdone by silver, a colour symbolising luxury and wealth.

The colour of a product influences the perception of it, and this is used extensively in marketing. Many people believe a red car will drive faster than a white one. Tests with coffee showed that coffee served in a red mug was preferred to the same coffee served in a yellow mug (considered too weak) and in a brown mug (too strong). Foodstuffs claiming to be pure and unadulterated often use blue and white packaging to communicate purity. However, in this age of environmental awareness it is predicted that more earthy colours will symbolise a natural product while white will be associated with chlorine and all that is environmentally unfriendly. Products communicating strength adopt vibrant and contrasting colours; the greater the contrast, the stronger the associated power.

Colour can be used to represent a product, with some colours idealising a product, and some using biological signals to communicate function. Yellow and black are used in nature as a warning symbol for reptiles and insects that have poisonous bites or stings and this colour combination is therefore often used to represent danger. Yellow and black is used on signs when the desire is to signal caution.

Colour mixing. There are two ways to mix colour — 'additive' and 'subtractive'.

Additive colour is the mixture of coloured lights. The three primary colours of red, green and blue when mixed together in equal colours produce white light. Mixing the three additive primaries in differing amounts of coloured light can create any colour in the rainbow. Colour televisions use the principle of additive colour mixing.

In subtractive colour mixing the principle is exactly the opposite. The subtractive primary colours are cyan, yellow and magenta. When mixed together they subtract from the light producing black. When different pairs of the subtractive primaries are mixed, the colours red, green and blue are produced. These principles are used when we paint and in photography and colour printing.

Colour systems. As the human eye can distinguish between 10 million colours, it is clear that to describe colour experiences by name is imprecise. While everyone knows what is meant by tomato red, different people have different ideas as to exactly what colours are meant by beige and sand; mauve and lilac. In order to accurately describe or pinpoint colours, a reference or colour system is required.

All the attempts to notate colour can be traced back to the work of Sir Isaac Newton. In 1660 he re-created a spectrum by directing a narrow beam of white light through a prism; he went on to develop a colour wheel by taking the two ends and bending the spectrum into a circle. This colour wheel evolved and changed over the centuries and in 1810 Otto Runge created a spherical model, with white at the North Pole and black at the South Pole and with Newton's colour circle forming its equator. In 1915 William Ostwald devised his double-cone colour solid; also in 1915, Albert Munsell developed another system of colour notation that added steps to the constituent hues.

Munsell allowed his three-dimensional colour solid to respond in shape to the different potential strengths between hues, creating an asymmetrical colour solid. His colour model can be explained in three dimensions, in terms of hue, value and chroma. *Hue* is the 'colour' of a colour, i.e. its redness, greenness or yellowness, *value* refers to the amount of lightness or darkness, while *chroma* refers to its saturation or its colour strength.

A number of commercial colour specification ranges have been developed from Munsell's work including the BS4800 and the Pantone and Colour Dimension systems. In textiles and clothing, the *Pantone* system is probably the most commonly used, and a whole range of products is available in its defined colours. These include ranges of marker pens, paper and even fabric samples.

Colour psychology. There is a commonly held belief that red is exciting and green calming. But which red and which green? Research has shown that strong greens, yellows and blues are all seen to be as exciting as strong red. The same research also showed that pale and dark versions of the same hues have quite the reverse effect. Pink has a calming effect on violent people. A common belief is that blue increases the size of a room while red has the opposite effect. Recent research, however, has shown that it is not the colour itself that has this effect but rather its value, the amount of whiteness or blackness. However, with regard to temperature, colour does have a very definite effect. Red and yellow rooms are perceived to be warmer than blue and green rooms.

4.1.1.6   Value (tone)
Between the whitest white and the blackest black there are countless degrees of light and dark values.

Value is important to the designer. Shapes that are close in value or tone appear to merge together. Visually dark values appear to come forward and light values tend to recede, but the reverse can occur. Sharply contrasting values attract attention and the use of light against dark or dark against light can create the illusion of size difference.

Value is probably the most elusive of all the design elements. The success or failure of a piece of work may rely on the use of values within it.

4.1.1.7   Texture
One definition of texture is the surface of a material as perceived by touch. Running a hand over a surface may find it to be smooth, rough, dull, glossy, hairy, sandy or bumpy. Texture is very important in textiles, and how fabric feels is an important consideration in product choice. Texture can be seen as well as touched and artists and designers can use a variety of techniques to convey textures when none is actually there. Repetition of design elements can often create a visual illusion of texture.

4.1.2   Introduction to design principles
In every design or work of art there are some or all of the design elements — varieties of line, positive and negative shape, three-dimensional form, occupied and unoccupied space, colour, value and texture. The manner in which these elements are used and combined determines the quality of a piece of work. Thoughtfully balancing, moving, repeating, emphasising and contrasting the design elements can achieve a unified piece of artwork or design.

4.1.2.1   Balance
Balance is a sense of stability when applied to opposing visual attractions or forces. There is a natural desire for balance, and in nature balance is ever-present.

In formal balance, the design elements are almost equally distributed. A design or composition that is divided in half so that one side is the mirror image of the other is said to have symmetrical balance. In radial balance, the design elements radiate from a central point as the spokes of a wheel or the natural form of a daisy.

Asymmetry uses informal balance; a centre line or point is ignored, with the design elements being balanced visually, rather than in a symmetrical manner.

The position of any particular shape in a composition contributes to its strength. A shape in the exact centre of a picture plane is at perfect equilibrium. Moving the same shape off-centre can increase or decrease its importance.

In any textile design, each shape affects everything else. Exciting balance or imbalance is usually only achieved after frequent arranging and rearranging. The study of composition has been a fascination for many artists and designers. The Pointillist artist Seurat was as equally absorbed in the study of composition as in the study of light.

4.1.2.2   Movement
Movement and the portrayal of movement have always fascinated artists and designers. By careful arrangement of the design elements, the illusion of movement can be created. In optical art and designs the sensation of movement may deeply affect the viewer's responses. Some paintings can provoke dizziness by making it difficult for the eyes to focus on a central point.

Associated with movement is time. Pictures and patterns are capable of holding our attention for varying amounts of time. Some designs may be so subtle that these are barely noticed, if at all, by the viewer, while others can hold attention for much longer periods.

4.1.2.3   Repetition
Repetition occurs when elements that have something in common are repeated. When a design consists of shapes that are exactly alike, repeated in a uniform and regular manner, then that design tends to seem more formal. By varying the shapes and the spaces between them, a more informal interest is created. The repetition of some of the elements within a design repeat can hold designs together.

Repeated shapes make patterns. Many textile designs, because of the method of manufacture, will automatically repeat.

4.1.2.4   Emphasis/contrast
Emphasis calls attention to important areas of design and subdues everything else on the picture plane. By placing emphasis on certain areas, artists and designers create centres of interest that cause our eyes to return there again and again.

Bold details, unusual textures and bright colours are more prominent than more subdued features. Often the left side and the upper part of a picture attract our attention first; this is particularly so for those whose language is written from left to right.

Similarity of elements in a design often leads to monotony. Contrasting elements tend to stand out. Elements that contrast strongly stand in opposition to one another — light against dark, large against small, round against square or smooth against rough.

4.1.2.5   Unity
Unity exists when all the elements in a design work together harmoniously. In a unified design, each element plays an equally important part.

4.2   Inspiration for textile designs

Inspiration for textile designs can come from a variety of sources. It is possible to create textile design work by going straight to the process being used. For example, a weave designer may go straight to the weaving shed, choose warp yarns in colours that appeal and, with a knowledge of fabric structures, make an attractive fabric. Experience, however, has shown that such a method of designing often results in fairly mundane design. Innovative, exciting textile design starts with much more fundamental paperwork; drawings and paintings exploring colours, textures, shapes and patterns. Drawing from objects in an imaginative and open way will inspire new colour combinations, textural ideas, shapes and arrangements of these.

Inspiration for such paperwork, and ultimately for textile designs, can come from many things. Both natural and man-made objects can inspire. Whether the source material is natural or man-made, abstract or tangible, how themes and source material are used is a matter for debate.

Drawing from man-made items such as architecture can raise the question as to whether or not the resulting design work is truly original. Drawing from architectural patterns can be seen, if the work is very literal, as merely copying someone else's designs — the proportions and shapes have been created by another person. Working from nature, however, is usually seen as producing more original work. The designer may still be copying but is constantly making decisions and judgements about what to include and what to omit.

Being influenced by the work of artists and other designers raises the question — when does a design influenced this way become a copy? Is a textile design that lifts elements from other designers or fine artists good or bad design?

Fine art can inspire textiles; part of a menswear collection by the fashion designer Paul Smith in the late eighties was made in fabrics that were inspired by the paintings of Matisse. Textiles can be created by fine artists; tapestries and silk scarves have been designed by artists and sculptors such as Picasso and Henry Moore. Textiles can have fine art directly applied; reproductions of the Mona Lisa have been applied to a variety of products, from tea towels to tee shirts.

Much of textile design is derivative. Source books for designers often consist of designs from the past or from other cultures. William Morris, the famous nineteenth-century writer and designer, often based his designs on Persian textile designs and many of the designs credited to him that are still best-sellers today were almost literal copies of the original Persian ones. Does this knowledge change how Morris's work is perceived? Does using archive or historical sources allow copying to become acceptable?

Whatever the inspiration or source for a textile design, the textile designer needs to have a good understanding of colour and aesthetics. A textile designer must be able to draw. They must understand and be able to use colour and pattern.

4.3   Pattern

Any textile designer needs to have a good idea of how their design is going to look as a finished fabric. One piece of paperwork could be worked through to become many different finished designs, woven, knitted or printed. All designers need to understand how designs can repeat. While many designs work as straight repeats, where the design element is repeated again and again, one on top of another, side by side, there are also half-drop repeats and tile repeats (see Fig. 4.1–4.3). Mirroring (Fig. 4.4), where elements are reflected about either, or both, x and y axes, is another common way of repeating elements.

Fabrics may have sections within one complete repeat that show other repeat patterns, while some fabrics such as border designs will employ repeated designs only within certain sections of the fabric.

It may be useful to mask-off certain areas of paperwork (or sample blankets and sample fabrics in the case of woven and knitted fabrics respectively), or use a frame to establish the most suitable sections to develop. It is important to consider the repeat structure from an early stage in the design development, and various sketch plans of possible repeat structures should be evolved.

Very often, trying to interpret paperwork too literally can be a problem, particularly for a print designer. By its very nature, weaving and knitting design requires elements to be extracted from the initial inspirational paperwork. For print designers, this decision-making process of what to include and what to discard is equally important. The best designs are so often simple. This advice is as true for printed textile design as for any other type of design. Everything that does not contribute, everything that is not necessary, should be left out.

4.3   Basic repeat structures

4.3.1.1   *Straight repeat*
A straight repeat is a simple repeat where the motif (in Fig. 4.1 an h) is repeated directly above and below in straight lines.

Fig. 4.1   Straight repeat.

4.3.1.2   *Half drop*
This time the design columns (vertical) slide halfway down, in a lengthways direction. Quarter drops and other fraction drops can also be used. (See Fig. 4.2.)

Fig. 4.2   Half drop.

4.3.1.3   *Tile (or brick) repeat*
Tile or brick repeat is another simple repeat where the motifs are repeated rather like a simple brick wall pattern. The second row slides halfway across in a widthways direction. (See Fig. 4.3.) The bottom row can also slide across other amounts.

Fig. 4.3   Tile (or brick) repeat.

4.3.1.4   *Repeat mirrored vertically and horizontally*

In this repeat the first motif is mirrored horizontally. This mirror and the first motif are then mirrored vertically to form a compound motif that is then repeated in straight lines (see Fig. 4.4). A wide variety of different repeat patterns can be created using mirroring.

Fig. 4.4   Repeat mirrored vertically and horizontally.

All the previously described repeat structures can be combined to provide a limitless range of designs.

4.3.2   The influence of end-use and methods of manufacture on repeat size

Factors which influence repeat size and the way repeats are constructed are end-use and method of manufacture. In determining a repeat structure for upholstery, the dimensions of the chairs and couches will have to be considered. Designing for duvet covers or headscarves also often means that the design is sized appropriately for the end-use. Such designs are said to be 'engineered'.

When designing for print, the method of printing will have an influence on the print size; design size for rotary screen printing or roller printing should be appropriate to the circumference of the screen or roller. The depth of the print repeat would normally be a fraction of the circumference, so that a whole number of repeats fit round the rotary screen or roller.

4.3.3   Centring

For the most economic fabric usage it is better for fabrics to be symmetrical in both the vertical and horizontal directions. This means that laying-up plans for cutting out pattern pieces can be more easily worked out. If a fabric has a pile, it is important that the pile lies the same way on all garment pieces.

Centring is when a fabric design is organised in such a way that it is balanced about the middle line of the fabric in a vertical direction.

4.3.3.1   *Example A*

Consider a striped fabric that consists of two colour stripes of equal width. The required finished fabric width is such that an exact number of the design repeat fits (see Fig. 4.5).

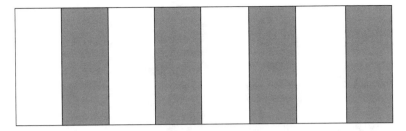

Fig. 4.5   Striped fabric not centred.

It would be better to have half a white stripe at either side as in Fig. 4.6 or a grey stripe at either side as in Fig. 4.7.

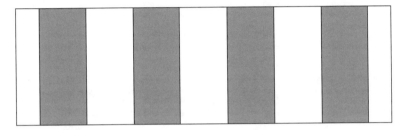

Fig. 4.6   Stripe centred with white stripe at either side.

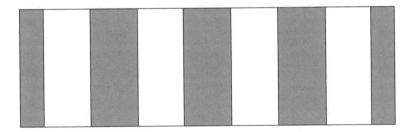

Fig. 4.7   Stripe centred with grey stripe at either side.

4.3.3.2   Example B

Consider a simple half-drop print design as in Fig. 4.8. This is not symmetrical about the vertical axis and so will give problems in cutting. Fig. 4.9 shows the same design centred and, as such, it can be more easily laid-up in layers, for cutting into individual pattern pieces.

4.4   Summary

Textile designers require to have a good understanding of aesthetics. There are what can be termed 'design elements' and 'principles' that are common to fine art and design. The design elements can be given as line, shape, form, space, colour, value and texture; the principles as balance, movement, repetition, emphasis/contrast and unity. Every design or work of art has some or all of the design elements, and some or all of the design

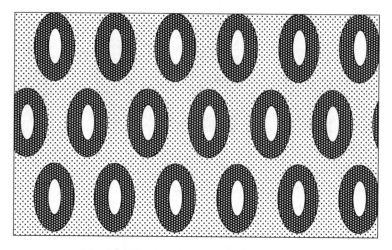

Fig. 4.8   Simple print design — not centred.

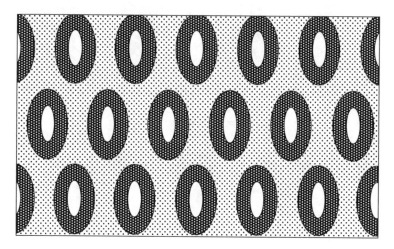

Fig. 4.9   Simple print design — centred.

principles are evident in the way elements are used. Inspiration for textile designs can come from a variety of different sources and this can raise questions as to what is original. Pattern and repetition are an integral part of most textile design, and how design elements repeat is very important to the textile designer. The most common repeats are straight repeats, dropped and tile repeats, borders, and repeats that use reflection and mirroring. The way fabrics are manufactured and used will have an influence on the way sections are repeated. The finished fabric width will have a bearing on the repeat size and how the design is laid out. Centring is where a design is repeated across a fabric width so that both halves are the same (or similar). This helps with laying-up and cutting out of the fabric pieces for making up into garments or other products.

References

1. Malcolm, D.C., *Design: Elements and Principles*, Worcester, Massachusetts, Davis Publications Inc., 1972.

Bibliography

Gregory, R.L., *Eye and Brain: The Psychology of Seeing*, 4th ed., London, Weidenfeld and Nicolson, 1990.

Home, C.E., *Geometric Symmetry in Patterns and Tilings*, Cambridge/Manchester, Woodhead/The Textile Institute, 2000.

Justema, W., *The Pleasures of Pattern*, New York, Van Nostrand Reinhold, 1982.

Maier, M., *Basic Principles of Design* [translated from the German by Joseph Finocchi], New York, Van Nostrand Reinhold, 1977.

Malcolm, D.C., *Design: Elements and Principles*, Worcester, Massachusetts, Davis Publications Inc., 1972.

Meller, S. and Elffers, J., *Textile Designs*, London, Thames and Hudson, 1991.

Phillips, P. and Bunce, G., *Repeat Patterns: A Manual for Designers, Artists and Architects*, London, Thames and Hudson, 1992.

Porter, T., *The Language of Colour*, Slough, ICI, 1990.

Rowland, K., *Pattern and Shape,* Vol.1, London, Ginn & Co., 1966.

Verity, E., *Colour Observed*, London, Macmillan, 1980.

5

Commercial aspects of design

There are many different aspects to design, including activities that are a result of the commercial nature of industrial design. For a design to be successful, it has to have a market, and, as most textile designs will reach their ultimate consumer via retail, designers should have some understanding of the retail trade and how this is structured.

5.1 The organisation and functions of a retail business

Even the simplest one-person business such as a small market stall involves the proprietor in all the functions of a retail business. Decisions have to be taken as to what is going to be stocked, stocks have to be ordered and deliveries have to be checked against orders. Advertising posters/flyers encouraging people to buy have to be made and displayed/distributed, the goods for sale have to be displayed, customers served and money taken. The money taken in payment has to be recorded and suppliers paid. Facilities for credit (with the necessary credit checks) and payment by instalment may be arranged, deliveries may be made to customers and, when business expands, staff have to be employed, trained and arrangements made to pay their salaries.

5.1.1 Buying and merchandising

The management need to know what and how much to buy. They need to know how to share out their money between the various sections or divisions of the business and then how much of each individual item to buy. When money allocations are calculated, the goods must be bought; this means buyers seeing manufacturers, representatives and agents, and visiting trade fairs to put together a suitable range of merchandise for their organisation. When items fail to sell, the merchandise section must decide when and by how much to reduce the price. In a large store or store group there will be both a buying and a merchandising section.

5.1.2 Receiving

When they arrive, goods are checked against original order and delivery note and examined for quality. Textile products are often checked against sealed samples. These are samples that have been checked and accepted for quality and retained by the retail organisation so that production quality is confirmed to be the same as sample quality.

5.1.3   Advertising and display
Goods are advertised for sale in newspapers and magazines. Window displays encourage customers into the store and in-store displays encourage maximum sales. Special offers and promotions can also be used to encourage sales.

5.1.4   Selling
This is the most obvious aspect of retailing where products are exchanged for money, or the promise of money in the case of credit card transactions.

5.1.5   Accounts
The accounts section deals with all the financial transactions, dealing with credit and paying suppliers. Cash must be accounted for and kept safely.

5.1.6   Personnel administration
Personnel deal with staff matters such as recruitment and training.

5.1.7   Salaries and pensions
This section is responsible for paying salaries and organising pension schemes.

5.1.8   Despatch
This section looks after the delivery of goods that have been purchased.

5.1.9   Maintenance and cleaning
The building and furnishings all have to be kept in a good state of repair and also clean and tidy. These functions will come under a section responsible for maintenance and cleaning.

5.2   Different types of retail structures

There are many different retail organisations; Table 5.1 gives a brief outline of the main types. Organisation charts for a multiple/variety chain and a department store are given in Figures 5.1 and 5.2 respectively.

5.3   Merchandise

The most important element of the retail mix according to most retailers is the product.

Table 5.1   Retail organisations.

Type of retail organisation	Main characteristics	Notes
Independent retailer	Owner/manager Often specialises Walk-in locations	
Co-operative society	Founded on co-operative principles, where members controlled organisation democratically and were paid dividends on purchases	
Department store	Selling under one roof, but in physically different departments, four or more classes of consumer goods, one of which is women's and girl's clothing	
Multiple/variety chain	Multiples are groups of speciality shops; variety chains are a group selling restricted ranges of different merchandise Strong corporate identity	Often concentrate on fast-moving lines Many target well-defined customer groups Centralised advertising, personnel recruitment and training, operating policies, etc.
Supermarket/hypermarket	Large range of items, covering foods and some non-foods Self-service In-store use of most advanced point-of-sale equipment	
Discount store	Very low prices Low profit margins Emphasis on own brands	
Franchise	Most normally for retail catering Franchiser supplies equipment and/or raw materials Strong corporate identity	

Mail-order via
  catalogues
  press adverts
  direct mail shots
	'Free' credit
	Wide selection of merchandise
	Customers' trial and approval facility
	Money-back guarantee
		Importance of merchandise being able to photograph well
		Colour reproducibility in catalogues is a big problem
		Returns can be high

e-mail	Global	Again problem with showing goods clearly

Markets
Wholesale
  Cash and carry
  Specialist wholesale
	Low prices
	Sells to trade
	Payment usually in cash
	Customer takes goods away	

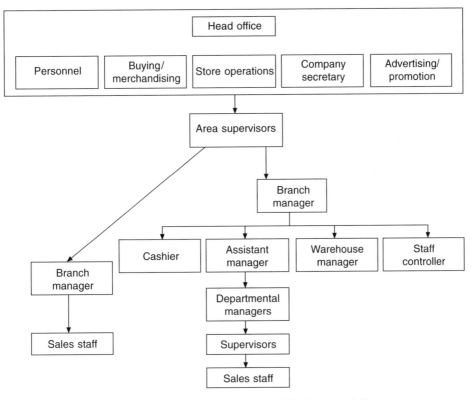

Fig. 5.1 Organisation chart of a multiple/variety chain.

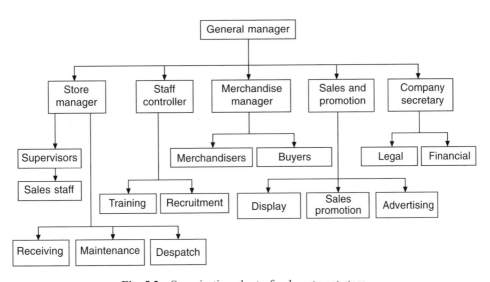

Fig. 5.2 Organisation chart of a department store.

In retailing, the merchandising process covers the selection, purchase, stock-management and display of a range of products whose successful sale will achieve the marketing objectives of the business. Merchandising is having the right goods of the right quality at the right time, in the right place, in the right quantity, at the right price.

5.3.1   Buying

There are five major areas to consider: who buys, what to buy, how much to buy, from whom to buy, when to buy.

Most textile and fashion products these days are not sold directly to the public but through mail-order or retail. The buyers and selectors, who are the people responsible for deciding which styles are going to be available, play a key role in what will become fashionable. While they cannot decide what the public will buy, they do set the limits within which the public have to make their choice. They act almost as a filter.

Buyers select from the styles those styles which they 'feel' will sell in their particular retail or mail-order outlets. Often buyers will influence styles by asking for changes to the original designs in terms of colour, fabric, trimmings, etc. Some merchanting companies employ designers to design ranges exclusive to them.

Buyers build-up knowledge of their customers, their section of the market, and they make their judgements about what to buy in the light of that knowledge. They also have to watch closely what is happening in the fashion world. They attend fashion shows, exhibitions and fairs. They talk to designers, manufacturers, other buyers, fashion consultants and journalists. By reading newspapers and magazines they see what their competitors are doing. In this way they get a feel for what they consider to be right for the next season. Like designers, they have to try and anticipate the way taste is moving. Buyers are powerful but they do make mistakes and they cannot force the customer to buy. In an organisation, one individual or a team of specialists may be responsible for the buying function.

A successful buyer should be able to: recognise customer needs, know the merchandise, be a good judge of quality, be a good judge of re-saleability, keep knowledge up to date, liaise effectively with suppliers, and plan effectively in numerate terms.

Different types of retail operation will have differently organised buying structures. Often, the buying functions within the different departments in department stores are separate. Multiples usually have central buying for the whole group while co-operative organisations may have several different buying structures with some centralised buying, some buying as a consortium, some inter-group buying and some branch buying.

Retail organisations can buy from different types of suppliers such as manufacturers and primary producers, wholesalers, importers, agents, other retailers, manufacturer-owned retail chains, government and semi-government sources and even the public. Decisions as to which supplier will be used will depend on several factors including price, terms, deliveries, service and suitability of product range.

The media are very influential in persuading the public to buy. Large and influential retailers employ the service of Public Relations companies to make sure their shops and goods are promoted on fashion pages of newspapers and magazines.

Journalists writing about fashion in clothes and furnishings are also part of the selection process. They select the work of certain designers and clothes from certain manufacturers or shops from the whole array of styles and products displayed at fashion shows, exhibitions and in the shops. The press cannot describe everything, so they draw attention to a selection of the styles available. In the process, they mark out certain styles as being fashionable.

For a long time, the press looked only to Paris and Milan, but now they include the collections of young designers all over the world. The press attempt to identify themes in terms of design inspiration, colours and styles, and most attention is given to designers whose work corresponds to the themes that journalists have identified. A similar process takes place with articles written for the textile and clothing industries' trade journals.

At fashion shows and exhibitions, press releases and publicity material also highlight themes; identifiable trends that other manufacturers can borrow, buyers can interpret for their stores and journalists can pass on to the public.

While not all buyers will want to follow these trends (they may be looking for a different image or they may feel the trends forecast are not suitable for their particular market), they will inevitably be influenced by the highlighted themes.

As well as having an important role in identifying trends and defining certain styles as fashionable, and in spreading that information within the manufacturing and retailing industries, the media also play an important role in disseminating fashion information to the general public. Most women get to know about new fashions by looking in the shops, seeing what other women are wearing, watching TV (a huge influence) and by reading newspapers and magazines. Window displays in shops show current stock, often displaying these in such a way as to create a specific mood or style. Purchases are made from a particular shop for a variety of reasons; for example, because the consumer has a desire for something they know that shop will stock or because they consider a particular shop fashionable.

The spread of fashion in society — from fashion restricted to the elite and wealthy, to fashion for all — can be associated with the growth of mass-manufactured fashion and also with the growth of the media, in particular women's magazines.

5.4   Information generation

Designers are at the starting point for new textile products and, as such, they are required to generate information for use by different departments within manufacturing and retail. Effective communication skills (visual, verbal and written) are essential for this.

The information designers generate may be in the form of reports, gathered for several different reasons; fabric specifications, to enable designs to be taken into production; and information for costing. Designers will also find themselves writing to customers and clients, both giving and asking for information.

5.4.1 Letters

Letters are written for a variety of purposes. For example, a designer may write to a customer to confirm instructions that they have been given or they may write a letter applying for a job. They may write a covering letter when they submit samples, explaining any instances where they have perhaps adapted design ideas for ease of production. And for freelance designers, the fee letter, where they set out terms of reference for a design project, can become a legal contract between designer and client (see Chapter 6).

A letter should be easy to read, pleasant, logical and clear. It should be as short as possible while covering the subject matter adequately. Success in communication ultimately depends on the reader's or listener's response. For a designer, as a person solving design problems, the layout of any letter is important.

See Appendix C for sample letter.

5.4.2 Reports

Reports are written to communicate information. There are several circumstances when designers may find it necessary to generate reports. Some of the main types of report required from designers are summarised below.

- *Research report.* Written record of research work carried out and the findings.
- *Design report.* A designer's work is largely visual but often some words of explanation are necessary.
- *Trend report.* Written and visual report of predicted future trends.
- *Market report.* Report of what is available in the marketplace with appropriate conclusions.
- *Consumer report.* Report on target customer profiles and lifestyles.
- *Report on trade shows/exhibitions.* Written or visual report to colleagues on what has been seen.
- *Customer/supplier visit report.* Written report to colleagues to keep them up to date with events.

Reports allow information to be digested at leisure. Authors of reports will be judged to some extent on their writing skills. The quality of a report is often taken as an indication of the quality of work as discussed in the report. Real communication means thinking more about the reader than self-satisfaction.

As with a letter, a report should be easy to read, pleasant, logical and clear. Again, it should be as short as possible while covering the subject matter adequately.

Designers are judged on the visual presentation of any report. If a designer cannot succeed in communicating visually in a report, how can they solve design problems? First impressions count. The appearance of a report has a good or bad effect on its readers before they have even begun to read it. The first part of a report to be seen is its cover; therefore suitable packaging is essential. The cover should make it clear who has written the report, what it is about, and when it was written and on whose behalf.

Reports should be complete in their own right. Any tables, references or other material needed should be included in appendices.

5.4.2.1   A basic structure for report writing
Abstract. Summary of the project in approximately 300 words.
Acknowledgements. Recognition of any help given.
Contents page.
Chapters. Covering — background,
discussion of concepts and theories,
research carried out by others,
methodology used,
summary of findings,
recommendations.
References. Any publications used should be included.
Appendices. Material relevant and referred to in the report.

5.4.2.2   References
Full details of all publications used should be given using a recognised citation system. There are various systems in common use and some examples are given below.

For a book: Author, title (in italics), place of publication, publisher, year. E.g. Harris, J., *5000 Years of Textiles*, London, British Museum Press, 1993.

For an article from a journal or magazine: Author, article title, journal title (in italics), year, volume (in bold), issue, page number(s).

For a web page: Author (if known), title of document (in quotation marks), title of complete work (if applicable, in italics or underlined), date of publication (if known), date of access, URL (in angle brackets). E.g. *John Smedley Web site*, February 2000, 18/6/2000,<http://www.johnsmedley.com>

5.4.2.3   Oral presentations

Sometimes it is desirable for designers to present their work in person, to talk through what they have done and why. Preparation for any such presentation is very important, to make sure that it is logical and clear. It is vital to ensure that everything that will be needed is there (and working), and running through everything in some detail before hand will pay off. Presentation skills, like anything else, improve with practice. The following guide summarises some of the key points to remember when making any oral presentation:

- Allow ten times as long for preparation as the presentation will take. A fifteen-minute presentation will take $2^1/_2$ hours (150 minutes —15 × 10) to prepare.
- Speak clearly, with varying speed and tone of voice.
- Focus on what your audience wants and will gain from your presentation.
- Maintain eye contact with your audience.
- Listen actively to what is said to you and respond positively.
- Be yourself. Be honest. Do not attempt to be an expert on everything. If you are asked questions to which you do not know the answer, say so, but offer to follow up by getting accurate answers — and do so.
- In presentations to groups, involve individuals wherever possible. This has the effect of drawing the group together.
- Be wary of using humour; don't poke fun at individuals or put yourself down.
- Watch body language. Keep feet still and leave arms free to reinforce points.
- Rehearse important presentations in front of a mirror.
- Run through your presentations to check the time these will take. Keep to the time allowed.
- Show confidence and enthusiasm.
- If you are using any visual aids,
    keep things simple,
    don't over-present so the presentation style overshadows the content,
    check, and double check for any grammar or spelling mistakes, and
    make sure any writing is easy to read from the back.

5.4.2.4   A basic plan for oral presentations

1) State the problem or opportunity as simply and concisely as possible.
2) Describe the negative effects — of the problem or of not taking the opportunity.
3) Give your recommendations/suggestions for action.
4) Describe the key features of your recommendations/suggestions.
5) Give the benefits of these features.
6) Give evidence to back the recommendations/suggestions.
7) Summarise the problem, the recommendations/suggestions and the benefits.

5.4.3   Fabric specifications

The design studio is where ranges are put together from fabric samples developed by the design team. The design manager will, with the designers, agree colours and fabrics. For production runs to be exactly the same as the agreed samples, the designers have to ensure that the Production Office has all the particulars required to make the fabric. This involves specifying exactly the colours, yarns, weave or knit construction, quality and finish for commercial production, and it is usually the designer who is the person responsible for passing initial production fabric against original samples before it is taken into full production.

Full making particulars are also required by the Production Office so that the quantities of yarn and other materials required to produce the fabric can be ordered. The weights of yarns and names of suppliers are required so that Production can order the right quantities and qualities of raw materials.

Usually fabric-making particulars will be recorded on a standard sheet. This will have several sections recording all the details required for repeating the sample in production.

While the making particulars would normally include the following, the exact particulars would, of course, depend on the type of fabric being made.

For all fabrics:

• A fabric name and/or number. While names are easier to remember, numbers can be less confusing. Frequently both may be used as identifiers of a particular fabric.

For woven fabrics (See Appendix D):

• A record of all the yarns used. The colour, count, quality and supplier of all yarns, together with the fibre content of each.
• The number of warp ends, the width of the fabric from the loom, and the finished width.
• Warp and weft plans with any repeats clearly indicated.
• The weave with suitable draft and lifting plan.
• The number of ends and picks per ten centimetres.

For knitted fabrics (See Appendix E):

• A record of all the yarns used. The colour, count, quality and supplier of all yarns, together with the fibre content of each.
• The fabric construction, suitably illustrated by loop diagrams or other notation system.
• Any striping patterns.
• The number of wales and courses per ten centimetres.

For printed fabrics:

• A record of the base fabric, including construction and fibre content.
• The dyestuffs and colour recipes used.
• A suitable record of the print with colourings.

For all fabrics:

• Any finishes that have to be applied (such as flame-retardant finishes, any brushing or raising, etc.).
• The weight of an appropriate measurement of fabric (grams per square metre or grams per finished running metre).

5.4.4   Percentage compositions

When several different yarns with different fibre compositions are combined in one fabric, the resulting percentage fibre composition for the new fabric has to be calculated. This will often be done by the Production Office, although sometimes it may be the job of the Design Studio. In either case, it is important for designers to understand the procedure. (See Appendix F.)

5.4.5   Information for costing

The making particulars as given on specification sheets, as well as being required by the Production Office for organising production runs, are also required for costing purposes. For a selling price to be determined for any fabric or textile product, the actual cost of making it has to be calculated. This cost price includes both direct and indirect costs.

5.4.5.1   Direct costs

Direct costs are those that can be directly attributed to a product or service. These can include raw materials, labour costs for weaving or knitting, any finishing cost (per metre, for example, for treating a fabric to impart flame retardancy) and distribution costs.

5.4.5.2   Indirect costs

Indirect costs are all those other costs associated with manufacture which cannot be directly attributed to any individual product or job. These might include the cost of sampling and design, the cost of the sales and marketing departments, rent and electricity. Such costs are also termed 'overheads' and an annual estimated figure for overheads should be linked into any costings, perhaps as a percentage or as a price per unit.

5.4.5.3   Profit

Profit is the monetary gain from being in business; the excess of revenues over outlays and expenses in a business enterprise. A percentage for profit would normally be incorporated into any costing exercise.

5.4.5.4   Selling price

The selling price for any product will normally be different from the cost price. At its simplest, the cost price is that price which essentially will cover any costs involved and ensure that no loss is made. However, determining an appropriate selling price has to take into account additional factors. What is the perceived value of the product? Selling an item at a very low price, while this may cover making costs and even allow a percentage for profit, may result in a perception that the product is of little value. Selling at the price as costed may also result in the product being in direct competition with a similar product line. Raising the selling price to take this into account may actually generate new sales without damaging sales of the cheaper product.

5.5   Summary

For a design to be successful it has to have a market, and as most designs reach this via retail it is helpful for designers to have some understanding of retail and the functions covered by retail organisations, particularly buying and merchandising. Buyers gather trend information to help them put together a range of goods for sale. They work closely with merchandisers who work out the quantities that should be ordered and, if an item is not selling, the merchandisers decide by how much it should be marked down.

A variety of information is generated by a design studio. Effective communication skills are therefore essential. Designers have to write letters and reports. These letters and reports are often about visits they make, about information they have gathered and about work they have done. Letters and reports should be clear and easy to read since, as visual people, designers are often judged on the way their letters and reports look. Fabric specifications allowing for accurate re-creation of fabrics are usually the responsibility of the design studio. These are also required by the Production Office for ordering the necessary materials and for costing. Designers may also find themselves working out the percentage composition of the new fabrics they create.

Bibliography

Cox, R., *Retail Management*, 4th ed., London, Financial Times/Prentice Hall, 1999.
Diamond, E., *Fashion Retailing*, New York, Delmar Publishers, 1993.
Leal, A.R., *Retailing*, London, Edward Arnold, 1974.
Packard, S., *Fashion Buying & Merchandising,* 2nd ed., New York, Fairchild Publications, 1983.
Sussmans, J.E., *How to Write Effective Reports*, Aldershot, Gower, 1983.

6

The professional practice of design – 1

Design covers many activities. Textile designers, in common with other designers, require to carry out these activities in a professional manner. While it is difficult to give a definition of professionalism, it is easier to state how professionalism is demonstrated.

Professionalism is demonstrated by

- self-confidence and flair,
- capability and expertise,
- rational and systematic thought,
- creativity and judgement,
- sensitivity to the environment, and to other professions, nationalities and cultures,
- appreciation of design practice within a global context,
- a lifetime commitment to personal education and professional advancement.

A professional approach is required for everything from answering job adverts and advertising services, through briefing meetings, presenting work, working out fees and writing invoices, to checking initial production runs.

6.1 Getting design jobs

Designers and design companies need to ensure that prospective clients or employers know of their existence. Potential clients need to know:

- who a designer/company is,
- what the designer/company can do (for them), and
- contact details.

Many freelance designers will already have contacts from previous experience in industry and prospective clients may already know of them and their work.

6.1.1 Advertising

Designers need to advertise themselves and their skills, whether working as a freelance designer or whether they want a job as an in-house designer. Well-designed and attractive publicity material is essential. Designers are visual people and whether filling-in an application form for a specific job, writing a CV, or preparing a brochure selling a design consultancy, these visual skills should be clearly recognisable. The written application letters, CVs and job applications of any designer must be well designed. A lot of money can be spent on publicity material and literature but, whatever the budget, presentation should be as professional as possible. Money spent on a logo and house style is money well spent as well-produced stationery is an enormous asset in terms of credibility.

6.1.1.1   Identifying potential clients/employers

Designers, when marketing their services or design work, need to identify potential clients/employers. There are several sources that can be used to make up lists of potential clients. These include:

- previous contacts,
- magazines,
- trade directories,
- catalogues from trade fairs and exhibitions, and
- information gleaned from other designers.

Once identified, the potential clients need somehow to be made aware of what is being offered:

- Existing and potential clients/potential employers can be contacted to advertise services offered.
- Publicity leaflets, CVs, and business cards can be printed and distributed.
- Trade exhibitions can be a platform for exhibiting work and making contacts. Such shows can be very useful but also can be expensive so should be used to full advantage.
- Information packs can be made up and given to the press covering the event. Such packs can also be distributed to any other parties visiting who may be interested in the services being offered.
- Local, regional and trade press, and telephone and other directories can be used for advertising services.
- Work can be sold through agents and, in this case, the agent will promote the designer's work for a percentage commission on any sales. As well as being used to sell work, some specialist agents will help with full-time employment.
- Public relations consultants can be used to promote services.
- Articles can be submitted to magazines and papers (see Appendix G).
- Names can be kept in front of existing and potential clients by informing them of any newsworthy achievements.

6.1.1.2   The Chartered Society of Designers code of practice concerning advertising

The Chartered Society of Designers (CSD) believes that it is in the interest of the design profession and of industry that the employment of qualified designers should be increased. It therefore suggests that members promote their own services and those of their profession, in a manner appropriate to the various fields in which they work. The CSD code also states that any claims made by a designer in their advertising must be factually correct, honourable and clear as to their origin, and that nothing should be advertised that might cause harm to a fellow member.

6.1.1.3   Job applications and CVs

Whether answering an advertisement for a position as an in-house designer or writing to advertise freelance services, textile designers need to have well-produced and well-presented CVs.

Job applications and CVs require information to be presented that covers several areas. These areas are normally:

- Personal information.
- Educational information.
- Employment history.
- Job-specific information.
- Transferable skills.
- Any other relevant information specific to a particular situation; this might include any awards received and membership of professional bodies.
- Hobbies and interests.

Personal information should include name, gender, date of birth, marital status, home and business addresses with relevant telephone numbers, fax numbers and email addresses.

Educational information would normally include schools and colleges attended, with dates and qualifications obtained. Employment history is probably better presented with the most recent position given first, working back in time through earlier positions held. Employers, positions held with responsibilities and dates should be given.

Any information specific to the job being applied for or type of freelance work being offered should be included. General transferable skills should be given, such as computer literacy, the holding of a current driving licence and any language skills (with degree of competency). There should also be a section on interests and hobbies.

Remember, as with any form of communication, a CV from a designer should communicate on a visual level. A well-designed CV shows that the person responsible is capable of solving design problems.

6.2   A model for design administration

Goslett, in her book *The Professional Practice of Design* [1], gives a model for design administration. This starts with the briefing and works through the design process to the completion of job with the filing of records. While it refers to designers working freelance or as consultants, many areas of this model are of importance to the in-house or staff designer.

Goslett's model comprises three phases:

Phase 1: *The fee contract.*
    (a) Being briefed.
    (b) Writing the fee letter.
    (c) After any immediate negotiations, receiving the written acceptance of this.
Phase 2: *Progressing the job.*
    (a) Setting up the job, research, preparation and submission of preliminary designs, followed by the first invoice.
    (b) Design development followed by subsequent invoice/s.
    (c) Finished working drawings, supervision of production, followed by subsequent invoice/s.
Phase 3: *Winding up.*
    (a) Publicising the job.

    (b) Final invoice.
    (c) Filing essential records.

The design business is diverse by its nature and there is no 'right' way to proceed. The important thing is to find the best way in any set of circumstances.

6.3   The initial meeting and briefing

The initial meeting with a prospective client/employer serves two purposes:

1 To convince the client/employer that the designer is the right person for the job.
2 To allow the designer to find out about the job.

It is worthwhile to find out as much as possible about a prospective client/employer before an initial meeting to discuss a project/job. The designer has to present himself or herself. It is important that the image given is professional. Designers should look the part, they should present a 'designerly' air and show themselves to be capable and organised. Designers should inspire confidence. It may be necessary to take a portfolio of work and, if so, the work should be carefully selected and presented (see Chapter 3).

    The meeting when the details about the work to be done are given is called the briefing meeting and it should be viewed as a two-way exercise enabling the client to state what they want, and the designer to clarify issues which spring immediately to mind.

    Wally Ollins, a design consultant, thinks that 'There is no such thing as perfect design, and good design is far too rare. In my experience, one of the major factors contributing to this is the inadequate design brief. The brief is the most important stage of the design programme. First, it should contain all the information necessary to enable the designer to quote for the work he is commissioned to do, and it will also be the basis of his contract with the client. Second, it must provide sufficient information for him (the designer) to make fully informed decisions in the initial stages of the design process.'

    The notes taken at a briefing meeting are very important. Once agreed in writing these will form a contract and protect both designer and client. Some clients know what they want while others haven't a clue, but most fall somewhere in between. A checklist can be helpful with regard to asking relevant questions. The designer needs to accumulate information.

- At what market area is the product aimed?
- Is it for home or export?
- What is the maximum production cost, selling price?
- Who are the main competitors at home and abroad?
- What is the time scale for sampling?
- For what season is the range aimed?
- Has the client any sales figure, reports, predictions or forecasts that could be helpful?
- Who are the customers?
- Are they mail order or retail?
- How big should the range be?
- How many fabrics, styles, colourways?
- Are specifications to be given?
- What are the production limitations?
- What manufacturing plant is available?

- Are there any restrictions to pattern areas?
- What are the making-up facilities? etc.

Often, the briefing meeting will take place at the client's manufacturing plant. If not, a visit to the manufacturing plant may be necessary to answer some of the questions posed.

A clear understanding of the aims and objectives of any project is fundamental to its success. A briefing meeting should allow a clear set of aims and objectives to be identified.

The in-house designer must also have a brief. This can be more informal and often the brief will not emerge from one specific meeting but will rather evolve and develop, with the understanding of what is required coming from knowledge collected over time. The in-house designer, however, still needs a clear idea of the aims and objects of any project.

6.4   Sizing up the job

After the briefing meeting, the designer should have enough information to be able to work out how he will set about solving the design problem. A programme of work can be devised for meeting the design brief. Experience will tell how long to allow for a job, and this will depend on the complexity of the job and how it will fit in with work already under way. Most jobs take longer than anticipated so it is important to ensure that enough time is allowed. Usually, any designer will be working on more than one job at any given time so deadlines should be realistic.

6.5   Agreeing terms of reference

Having considered what a particular job will entail and how long it will take to complete, for freelance and consultant designers the next stage is for both parties (designer and client) to draw up and sign one or more documents that will form the terms of reference for the job. This should be done before any design work is undertaken. These documents should set out all the factors relating to the brief, including the stages that the design will pass through and the fee for each stage, together with time scales.

The most usual form for such documents is a fee letter. This is a letter from the designer to the client confirming clearly the brief as it is understood, just what is going to be done and by when, what the fees are going to be and how these will be charged. The fee letter will also cover what will happen if the project is abandoned and how any work that becomes necessary, in addition to that initially anticipated, will be charged. There may well be mention of who will hold copyright of the design work done.

A well-constructed fee letter will always be impressive. Facts should be set out clearly and in the proper sequence since this letter, with the client's signed acceptance, will eventually become a contract between designer and client. It is therefore well worth spending time and trouble over each fee letter. (See Appendices H and I for a suggested structure, an example and more information on fee letters.)

The fee letter is the basis of all design work that is to be done. It reduces misunderstandings, making it clear just what is to be done and how it will be done. A real benefit for designers is that having to write a fee letter forces them into focusing on the project. It should be easier to determine just what needs to be done and therefore how much time they need to spend on it.

6.6   Fees — how much to charge

For designers working for themselves without the benefit of a regular salary, the key to successful fee estimating is knowing the selling cost of an hour of time. As well as the time taken to produce a finished piece of work, whether that time is spent in thinking, drawing, travelling or research, any designer will be selling their talents, training and expertise.

To work out the cost of one hour of their time, designers have to consider several factors. The basic cost to be considered is the personal budget or salary that is to be paid. The number of hours that can be sold also has to be considered. Holidays, time off for unforeseen problems and illness have to be taken into account. And it should be remembered that a lot of work chasing after potential jobs, exhibiting at trade shows and the like, is done speculatively and cannot be charged directly to specific jobs. Starting-up expenses and any capital expenditure also need to be taken into account and a percentage for profit will normally be added. A simple equation to work out an appropriate hourly rate would be

$$\text{cost of one hour} = \frac{\text{salary} + \text{overheads}}{\text{hours that can be sold}} + \% \text{ for capital outlay} + \% \text{ profit}$$

(See Appendix J for more details as to how an hourly rate might be worked out.)

6.7   Different types of fees

There are different ways of charging for design services. The hourly rate can be used in some situations but, more usually, fixed fees are charged and these should be agreed before any work is started.

6.7.1   Fixed fees

These are now the rule rather than the exception. Most clients will want to know exactly what the fees for each stage of a project will be before they commit themselves. The calculation of a fixed fee is relatively straightforward once it has been worked out exactly what the job entails.

Rather than wait until the end of a long job before sending an invoice, it is much better to invoice at the completion of each stage. If this is going to be done, then it should be indicated in the fee letter.

6.7.2   Hourly rates

Hourly rates are usually used when the designer is required to do extra work over and above the agreed amount. These fees would normally be invoiced monthly and would have been referred to in the fee letter.

6.7.3 Retainers

A designer may be asked to give a client regular input on design matters over a fixed period, usually a year. The fee would be worked out by negotiation with the client and would normally be invoiced on a monthly basis.

6.7.4 Royalties

Royalties are usually paid as a percentage of the actual sales of a product. The percentage agreed is based on the estimated sales so that if the product sells very well the designer shares in the profits. On the other hand, if sales are much lower than expected the designer loses out. Royalty fees are often combined with other forms of fee payment to give part fixed fees and part royalty.

6.7.5 Exclusivity

This form of contract is normally negotiated for a minimum period of a year. The designer is contracted to work on an exclusive basis in a defined product area. Fees should compensate for other work that has to be turned down.

6.8 Keeping records

For both freelance and in-house designers, a method of recording what is done must be established. In addition, the freelance and consultant designer must establish the time spent on any job and keep account of any expenses that might be directly attributable. In some cases perhaps it will be enough simply to give each job its own file. It may, however, be more useful to give each job or project a number. In any case, all paperwork, design work and fabric samples should be filed regularly in a reasonable system which will allow easy retrieval when necessary. It is worth spending time putting in place suitable filing systems. A lot of time can be wasted if there is no effective system in place for retrieving information.

6.9 Invoicing

For the freelance and consultant designer, when the design work is completed and has been presented to the client the invoice can be calculated and submitted. Designs could be submitted on a Monday and the invoice on the Tuesday; the invoice could even be enclosed with the work. This is, however, likely to create a bad impression. As a general guideline, an invoice for work submitted in the first three weeks of a month could reasonably be sent on the last day of that calendar month, while for work submitted in the last week it might be better to wait until the next month.

If designs are submitted and are waiting to be approved by, say, the board of directors, then it is usual to wait for this before sending the invoice. It is always worthwhile finding out whether or not the client is happy with work before submitting the invoice. If it states in the fee letter that fees will be invoiced monthly, then invoices can be sent monthly regardless of the stage reached. Final invoices must include everything that can be charged. It is only when all costs are known that the final invoice should be sent.

Invoices from suppliers are received monthly. These are called purchase invoices. Those sent out for design work done are sales invoices. Sales invoices should be sent out monthly, usually a week to ten days after the end of the month. An invoice sent on the 10th August would usually be dated 31st July or even simply July. This ensures that the receiving company's accounts department deals with it with their July invoices. If it were dated 10th August, it would wait until all the invoices for August had been received.

Invoices received will be checked against the order issued or the fee letter. Even if it tallies exactly, it will still probably have to go before the actual person who approved the work. This all takes time. With companies who have understaffed or inefficient accounts departments this can take a long time.

Some companies do not now pay on receiving an invoice; rather their policy is to wait until they receive a statement. If an invoice is not paid within a month, then it is usual to follow it up with a statement a month later. On receiving this it would be checked against the ledger entry and, in due course, a cheque would be made out, signed and countersigned before being sent.

This is why it is not reasonable to send an invoice one day and expect to be paid the next. To avoid problems with cash flow, it is important to invoice regularly and to ensure that invoices are as accurate and explanatory as possible to avoid delays.

6.9.1   Setting out invoices

An invoice will usually be sent on letter-headed paper. At least one copy should be taken as a record. The end of the month date should go at the top, with the name of the company it was being sent to. Invoices should never be addressed to a person; rather, they should be addressed impersonally to the company. Centred below this should go the word 'invoice'. A reference number should go next, followed by a brief general description of the job. Each item or sub-item ending in an amount of money should be numbered or lettered for ease of reference. A precise reference to a fee letter with dates and names should be made, if appropriate.

6.9.2   Statements

If an invoice is not paid within the calendar month after it was issued then it should be followed up by a statement. Statements should follow at monthly intervals until payment is made. As with invoices, statements are normally on letter-headed paper. The word 'statement' should be clear and the invoices outstanding should be referred to by invoice number and date sent.

Raising and chasing invoices is a very important monthly activity and time should be spent on this. Invoices and statements that are clear and do not raise queries will be paid more quickly.

6.10   Summary

Design covers many different activities and these activities should be carried out in a professional manner. Designers need to make sure that potential clients or employers know that they exist. This can be done by advertising and by identifying potential clients/ employers and making them aware of what services are offered. Job applications and

CVs need to be well produced and attractively presented. Goslett's model for design administration works through the design process from the briefing meeting to the completion of a job with the filing of the records. The briefing meeting is often the initial meeting where clear aims and objectives need to be identified. Once briefed, designers need to work out how they will go about solving the specific design problem. For freelance and consultant designers some kind of terms of reference need to be drawn up so that they and their client know exactly what is intended to be done. Often terms of reference will be set out in a fee letter; a letter from designer to client confirming the brief, setting out what is going to be done, what the fees are going to be and when the project will be completed. To charge appropriate fees, designers have to have some means of working out the cost of an hour of their time. Several factors have to be considered, including a basic salary, time for holidays and illness, time spent speculatively chasing jobs, any starting up costs and capital outlay. From a basic cost for an hour of their time, appropriate fee levels can be worked out. Fixed fees, retainers and royalties are different types of fees that can be charged; just what type is appropriate will depend on the job being undertaken. Effective record systems need to be put in place for paperwork, design work and fabric samples. For the freelance and consultant designers, when design work is completed, final invoices can be sent.

References

1 Goslett, D., *The Professional Practice of Design*, 3rd rev. ed., London, Batsford, 1984.

Bibliography

Goslett, D., *The Professional Practice of Design*, 3rd rev. ed., London, Batsford, 1984.
Lydiate, L. (ed.), *Professional Practice in Design Consultancy*, A Design Business Association Guide, London, Design Council, 1992.
Potter, N., *What is a Designer? Things, Places, Messages*, 3rd rev. ed., London, Hyphen, 1989.

7

The professional practice of design – 2

There are many organisations, both national and international, that support the textile design function. There are professional bodies that set out codes of conduct for their members as well as offering a variety of services, and there are trade associations that promote the products of their membership. Other organisations such as market research bureaux and trend publishers (see Chapter 8) offer important facilities.

7.1 Professional bodies

7.1.1 The Textile Institute
The Textile Institute is a worldwide professional association for people working with fibres and fabrics, clothing and footwear, and interior and technical textiles. It aims 'to set professional standards, advance knowledge and industrial practice, provide an interactive network for the exchange of ideas and create a social community promoting friendship among members'.

7.1.2 International Council of Societies of Industrial Design
Established in 1957 to advance the discipline of industrial design at the international level, the International Council of Societies of Industrial Design (ICSID) is a non-profit, non-governmental organisation supported by professional, promotional, educational, associate and corporate member societies on all continents.
   A member society of this organisation is the Design Council.

7.1.3 The Design Council
The Design Council is the UK's national authority on design. Its role is to advise and influence businesses and others on the importance of design in terms of its contribution to the United Kingdom's international competitiveness. As well as helping business and education, it aims to influence government policy making. It does this through high-profile events, initiatives for managing and implementing effective design, and specially-

commissioned TV programmes, as well as general media coverage of Design Council activities and publications such as policy proposals and specially-commissioned reports.

7.1.4 The Chartered Society of Designers

While the Chartered Society of Designers (CSD) is a UK organisation, there are similar bodies in many countries. As a professional body representing designers from all areas of design, the CSD's aims are 'to promote high standards in design, foster professionalism and emphasise designers' responsibility to society, clients and other designers'.

The society's activities include a comprehensive programme of training, including design management, marketing design and freelancing, programmes of events (social and professional), information services and advisory services on such issues as copyright and design protection, and publications including a newsletter and professional practice guides. It also represents members' interests to government, and gives general advice to companies.

7.1.5 The Design Research Society

The Design Research Society has an international membership and was established in recognition of the fact that design was common to many disciplines. There was also 'a recognition of the growing complexity of the social and technical worlds in which the designer works, coupled with a desire to explore socially-responsible, explicit, and reliable design procedures'.

7.1.6 The Design Management Institute

The Design Management Institute (DMI) has a vision to 'inspire the best management of the design process in organisations world-wide' and its mission is to 'be the international authority and advocate on design management'.

7.1.7 Services and opportunities normally provided by professional bodies

Such professional organisations as have been mentioned offer a variety of services and opportunities, which will normally include:

- Facilitation of networking between members.
- Setting of professional standards.
- Promotion of professionalism through a code of practice for members.
- Education and career development opportunities through exams and workshops.
- Recognition of talent and contribution.
- Dissemination of research and development through conferences and journals.
- Promotion of members as consultants.
- Help with employment.

7.2 Trade organisations and associations

As well as professional bodies, there are many different organisations within the textile and clothing industry that exist to help trade by promoting the sales of a specific product or range of products. Often funded by charges for membership, such bodies will market the products of their members by advertising, by producing (usually co-ordinated) promotional materials, and by putting on trade shows and other such high-profile events.

7.2.1   The Woolmark Company

This organisation exists to increase demand for wool, world-wide. Wool producers subscribe to have access to information and to have their wool promoted. The research and development unit develops new products and it provides the wool textile industry with a wide range of services. The development centre at Ilkley in Yorkshire integrates design, styling, marketing and technical activities, and represents a comprehensive resource for the wool textile industry.

7.2.2   The British Knitting and Clothing Export Council

The British Knitting and Clothing Export Council (BKCEC) aims to improve the export performance of the apparel/fashion industry by assisting both existing and potential exporting companies with marketing and promotional activities and by creating a greater overseas awareness of the British fashion industry.

7.2.3   The Knitting Industries' Federation

The Knitting Industries' Federation (KIF) is the national employers' organisation for the UK knitting industries. The services it offers member companies include help with industrial relations such as advice on employment law and representation at Industrial Tribunals, advice on environmental issues such as recycling and eco-labelling, health and safety, business mentoring schemes, advice on design and trend forecasting, and access to statistics on the industry. It also lobbies Government.

7.3   Business organisations

Design as an activity takes place within a business organisation. This organisation may be a one-person business or it may be a huge, multinational corporation. Whatever the case, designers should have an understanding of the different types of business organisations that exist. Many designers will, at some point in their career, consider working for themselves and some will actually take the plunge and set up their own business.

Before any business is set up, there are several things that have to be considered. These include financial implications, security, worry, holidays or lack of, length of working day, etc. For any designer considering setting up their own business, it is much better to research the market from the safety of a job where contacts can be made and potential customers identified. It is very difficult to set up in business straight from college or university.

It is essential when starting up any business venture to thoroughly understand the market and know which section of the market it is intended to serve. A survey of the marketplace and a profile of potential customers are essential; not only will this help with marketing strategies later on, but useful information will be picked up on the way.

A further consideration is where the work is going to take place. A studio at home can lead to distractions while separate business premises can be expensive.

7.3.1   Business structures

A business can be set up legally in several forms and each form has advantages and disadvantages. The laws governing businesses differ from country to country and professional advice should be sought before the decision to launch into business is taken.

In the UK, there are three main types of business structure: sole trader; partnership; and limited company.

7.3.1.1   Sole trader

This is when one individual sets up in business for themselves and it is the simplest way to trade. There are no legal formalities; basically all that has to be done is to tell the Inland Revenue and the Department of Health and Social Security the change in status. The business owner has independence and full personal control. However, the business owner is liable for all debts and hours can be long.

7.3.1.2   Partnership

Again there are no required legal formalities. A partnership is similar to a sole trader business but with the control and liability being shared between two or more. To avoid any possible future problems that may arise as partners' requirements change, there should be a legal partnership document drawn up. A limited liability partnership is a form of partnership agreement that, as its name suggests, limits liability in certain circumstances.

7.3.1.3   Limited company

A limited company has its own separate legal identity, with the directors being employees of the company. To set up a limited company there must be two shareholders, at least one director and a company secretary. If a company runs into problems, creditors can only take action against the company, not against the directors. Banks, however, will normally ask for some sort of guarantee against overdraft facilities.

Companies can be bought 'off the shelf', although these can be relatively costly to establish. To open a company bank account, a certificate of incorporation and a Memorandum of Articles of Association is needed. Trading is only allowed after incorporation, and audited accounts must be filed with Companies House.

Whatever the type of business, advice will be needed from an accountant, a solicitor and a bank. An accountant helps establish and maintain a business records system for income and expenditure. A solicitor assists with all matters legal, registration of the business, employment law, and health and safety legislation. A bank, as well as holding the necessary bank accounts, will be central to any business.

The support of a lending bank is fundamental to the growth of any business. A bank is in the business of lending other people's money and its first concern is profitability. A bank may share some risk but a banker will seek to ensure that the risks of any business are fully shared by the proprietors and the lenders. A bank will expect some personal commitment from the stakeholders in any business. In the early stages of any business, most finance is in the terms of an overdraft. The worst time to go to a bank and ask to borrow money is when it is desperately needed.

No one will lend money, or invest in any business organisation, unless it can be clearly shown what the money is needed for — how it is going to help and how it is going to be repaid. If a good case can be made then it is often not so difficult to raise money. If it is difficult, then perhaps the idea was not so good.

7.3.2 The business plan

Every new enterprise needs a business plan. This will set out what is intended, how the services are going to be marketed and sold, potential customers, and projected sales. It will also contain a profit forecast and cash-flow projection. If a case cannot be supported by projected profit and loss accounts, then there is no point in starting the venture.

Any business plan needs to include budgets for salary, capital outlay (starting-up expenses) and running expenses (some of which will be known, others of which will have to be estimated). Management ability, planning for contingencies and anticipating what might happen, will be reflected in the financial picture of the plan. Evidence of past performance might be given, or at least an indication of how things might be done. Writing a business plan gives an opportunity to take trouble to think ideas through thoroughly.

The most important features of a business plan should be:

- Main proposed activities.
- Resources required in terms of assets and manpower.
- Key people and how the business will be structured.
- Product or service with an indication of any unique or strong selling points.
- Product costings and break-even calculations.
- Profit and loss projections.
- Short description of any manufacturing process.
- Some indication of potential sales.
- Markets and strategy in terms of location and type of customers (brief details of market research should be given).
- Financial requirements.
- Training (with special attention to Health and Safety).

7.4 Legal protection

The legal situation with regard to design ownership is complex and, in cases of dispute, expert advice is essential. Some understanding of the basic principles underlying copyright and other design protection law however can be invaluable.

Unscrupulous businesses can make money by copying ideas taken from other companies' designs. Sometimes the copying company will re-interpret original design ideas, changing certain parts of the original design; sometimes the copy will be deliberately made so that it will be mistaken for, or 'passed off' as, an original. This 'passing off' of a copy as the genuine article often happens when products become well-known. Designer labels are constantly being copied and mimicked as cheaper copies have a ready market and can be very profitable for the manufacturers. While laws do exist which make any such copying unlawful, in practice it can be very difficult to take legal action against a copying company. The designer who has been copied has to prove that they were the originator and, in the case of 'passing off', that the copying company intended the public to think the designs or products were originated by the original designer. Action through the courts can be very expensive and it is often only larger companies that can pursue such a route.

7.4.1   Copyright

Although in practice few prosecutions are made for copyright infringement, a designer should still have a good understanding of copyright issues, both as to recourse available through the legal system when they find that their own work has been copied and to avoid inadvertent infringement of someone else's copyright.

While different countries have different laws, all countries have some copyright legislation. Essentially copyright laws protect design work from being copied. The work, however, must exist as a physical object; ideas cannot be protected by copyright. The work must also be original; it cannot itself be copied, even if the person who made what it was copied from is unknown or has been dead for a long time. Providing the work is original, copyright is automatic for a UK-based designer; the work should carry the international © symbol, the designer's name and year of authorship.

In the UK, any original textile design made after 1989 (when the latest copyright act was passed in the UK) is protected for the author's life plus 50 years after death. Copyright can be transferred to some other person (or body) in a written document, or it can be sold. In textile design there would be an assumption that full copyright was transferred to the buyer of a design unless otherwise stated. A textile designer who therefore wanted to ensure that their design for curtaining was not used for any other end-use (for example wallpaper) would have to ensure that they made it clear, in writing, that they were selling their design with restricted copyright.

7.4.1.1   Licence

Licence is the legal term for permission. Any person or organisation wishing to use a design which has copyright, can apply to the copyright owner for permission to use the design. This permission is given via a licence, which is bought by the person or company wishing to use the design. Most cartoon characters can be reproduced only under licence. Branded products which are protected (for example the cotton/wool blend 2×2 twill fabric 'Viyella') can also be produced only under licence.

7.4.1.2   Breaches of right

Copying a design that is protected by copyright breaks the law. Civil proceedings can be brought by the first owner of copyright of the infringed work, by a second or subsequent owner to whom copyright has been assigned, or by the person or company to whom an exclusive licence has been granted by the copyright owner.

Rulings against someone judged to have infringed copyright can include an injunction stopping the guilty party producing more goods infringing copyright, payment to the original designer of profits made by the infringer, delivery of infringing copies, artwork and equipment to the original designer, and compensation payments.

Companies using design require a working knowledge and understanding of copyright law, especially the rules about origination, first ownership, transfer and licensing. Clients need to know who will own copyright in any additional design work. Designers need to know that they cannot claim copyright of any of their own design work that involves the work of other artists or designers, even when that work is not copyrighted. They also need to be aware that 'borrowing' any imagery that is still protected by copyright means that they are in breach of copyright law and liable to prosecution.

To help in any dispute that might arise, original works should always carry the © by-line, author's name and date of creation throughout the development of the designs, on finished artwork and any merchandised articles. It is also advisable to clarify written terms and conditions of employment in relation to copyright.

7.4.2 Design right

Design right is intended to protect original design for goods, products and packages against copying, without the need to prove a breach of copyright. As with copyright, for work to be protected by design right it must be original. Design right does not protect one-off designs or three-dimensional works; it protects only two-dimensional design work that has been done for commercially or industrially produced items.

Like copyright, design right is automatic in UK law. No formality such as registration is required. Design right gives exclusive control over manufacture and distribution of the designed products to the person or organisation owning the design right. In work that has not been commissioned, the designer would own the design right; in the case of a commissioned design, the owner would be the commissioning client but the designer would still own copyright, although this would be reduced to 25 years. Normally design right would not go to an in-house designer; it would automatically go to their employer unless something different was stated in their contract of employment.

Design right is the best form of protection for designs that are mainly functional, and only incidentally artistic, such as clothing designs. Surface pattern design, however, is not protected by design right.

7.4.3 Length of protection

Works are protected by design right for five years before a product is marketed. From the time items are made available in the marketplace, the right runs for a further five years, then for a further five years during which time the design right owner must give licences to anyone prepared to pay a licence fee.

7.4.4 Design registration

Decorative and ornamental designs and designs automatically protected by copyright can also be registered. Designs to be registered must have been previously unpublished in the UK, and these should be mainly aesthetic, rather than functional. Protection lasts for five years after first registration and this can be renewed four times, up to a maximum of 25 years. The right to register is with the designer or employer if the design has not been commissioned. With commissioned designs, the client automatically acquires the right to register. Designs are registered with the Design Registry and, once registered, are given a registration number that is normally marked on the product together with the words 'Registered Design'. Textiles, designs for textiles, and patterns for coverings of walls and floors are often registered.

7.4.5 Patents

Patents offer protection for inventions, new products, and methods of manufacture. While artistic or aesthetic works cannot be covered by patents, novel methods of constructing or colouring fabrics may be.

7.4.6 Trade and service marks

There is a scheme for registering symbols or words which identify a particular type of goods or service. The aim is to protect brand identity against competitors and to achieve a monopoly for the name or logo in the marketplace.

A trademark is 'either a word, phrase, symbol or design, or combination of words, phrases, symbols or designs, which identifies and distinguishes the source of the goods or services of one party from those of others'. A service mark is the same as a trademark except that it identifies and distinguishes the source of a service rather than a product. Registered marks may be transferred or licensed.

A trademark is different from a copyright or a patent. A copyright protects an original artistic or literary work; and a patent protects an invention.

7.5 Summary

The textile design function is supported by many organisations both at a national and international level. There are professional bodies that advise and support their membership as well as setting out codes of conduct within which their members should operate, and there are trade associations which promote the businesses and products of their membership. Both types of organisation will lobby government on behalf of their members.

Design as an activity takes place within business organisations, and these organisations range from one-person businesses through to huge, multinational corporations. It is therefore helpful for designers to have some understanding of the different business organisations that exist and for which they might find themselves working.

Designers often turn to copyright law when they, their products or businesses are imitated. It can, however, often be difficult to establish a visual connection for copyright, design right or moral right purposes in the courts. The legal situation with regard to design ownership is complex and, in cases of dispute, expert advice is essential.

Bibliography

Goslett, D., *The Professional Practice of Design*, 3rd rev. ed., London, Batsford, 1984.

Lydiate, L. (ed.), *Professional Practice in Design Consultancy*, A Design Business Association Guide, London, Design Council, 1992.

Potter, N., *What is a Designer? Things, Places, Messages*, 3rd rev. ed., London, Hyphen, 1989.

Whitehead, G., *Business and Administrative Organisation Made Simple*, 2nd rev. ed., London, Heinemann, 1981.

8

Designing for the future

All designers are designing for the future. For designers to predict what people are going to want, they have to have a good understanding of the influences that affect desire for and ultimate choice of any product.

8.1 Purchase decisions

The factors influencing our choice of one product from another are complex and involved. There are aesthetic reasons for a purchase being made; to look good and enhance appearance. There are cultural influences, with purchases being affected by the buyer's culture and traditions, the prevailing attitudes and laws. Whether or not a product is within the purchaser's grasp financially or indeed whether or not it may be readily available as a result of production and distribution methods, also has an effect on customer choice. In addition, textile and clothing purchases are made to show status, with branded and designer-label products dependent on creating an image and appealing to a lifestyle. People are identified by the clothes they wear and the products they use.

8.1.1 Consumer buying behaviour

The buying process is essentially a decision process designed to provide solutions to problems. The first stage is the awareness of a need or want; this is followed by an information-processing stage prior to the purchase decision. The buying process is completed once the purchase of the product/service has been made. (See Fig. 8.1.)

8.1.1.1 The felt need/want
Human beings have certain basic needs for food, water, shelter and security. However, they also have other needs and desires that, while not being fundamental for survival, are still very important for well-being. There is a need to be loved, recognised and valued. Most humans have a need to conform; to be accepted. Generally, people feel comfortable in the company of people similar to themselves. Teenagers adopt their own uniforms outside school, dressing like those they wish to be associated with.

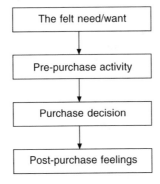

Fig. 8.1   Consumer buying behaviour.

8.1.1.2   Pre-purchase activity

This involves the gathering of information, the processing of which allows the consumer to move towards a purchase decision. Fig. 8.2 outlines the stages in pre-purchase activity. These stages are often summarised by the acronym AIDA — Awareness, Interest, Desire and Action.

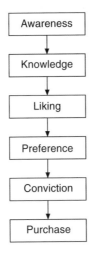

Fig. 8.2   Pre-purchase activity.

8.1.1.3   The purchase decision

A product is purchased to satisfy a need. What is the product to be used for? Who is going to use the product? When and where is it used? What other products is it used with?

The actual commitment to buy is never one decision but rather a bundle of decisions relating to product, brand, store and price. Ease of purchase, credit facilities, lack of queues, etc. will all affect purchase decisions, and additional in-store information can lead to trading-up and additional purchases being made.

8.1.1.4   Post-purchase feelings

After buying an item, there can be a variety of feelings about the decision to purchase. There can be satisfaction with a product that lives up to expectations or dissatisfaction through product failure. There are psychological factors to be considered; tensions created by wondering if the right choice has been made are common and this is the reason why most retailers have liberal returns policy. It is felt to be better to have a high rate of returns and keep customer loyalty rather than have few returns but alienate customers so that they will go elsewhere to make their next purchases.

8.1.2   Factors influencing product choice

The reasons as to why people choose particular products play a significant part in management decisions relating to what is to be offered, and the adoption by the buying public of certain products as fashionable. For the buying public to adopt any product or style, that product or style must be available and advertised as such. What is available on the high street is the result of a series of exercises in selection:

* selection by manufacturers as to what initial design ideas to pursue through to a finished product, at every stage — yarns, fabrics, garments and other textile products,
* selection by retailers as to what items to include in their product ranges.

What the consumer ultimately decides to purchase is influenced by further selection exercises:

* further selection by retailers as to what products and styles within their ranges warrant special attention and promotion through in-store promotions and window displays.
* selection by the media, press and television as to what will be promoted both formally through specific directional articles and programmes, and informally through articles and programmes where products are seen to be used by those taking part.
* finally there is, of course, selection by the consumers themselves.

The factors leading to the decision to purchase one item in preference to another are complex but will include availability, affordability, desire and need. These factors themselves are influenced by other factors such as technology, geography, income, status and culture.

There are several models that try to make sense of all the factors influencing product choice. Fig. 8.3 shows a model that divides the influencing factors into external and internal.

A second model, Fig. 8.4, groups these factors into six categories. It should be noted that all of these categories are inter-linked and inter-related. The categories are aesthetic, cultural and traditional, social–psychological, economic and political, managerial, and physical.

8.1.2.1   Aesthetic factors

This category includes factors such as creativity, taste and what is considered to be attractive or beautiful at a particular time. The desire to decorate and improve our appearance could also be considered within this grouping. These factors clearly influence product choice.

8.1.2.2   Cultural and traditional factors

Culture includes all those things that make up the framework of society, and that influence the adoption of new styles and products. In the mid-80s a designer fashion collection showed skirts for men; the prevailing attitudes were that such skirts looked odd and the style was unsuccessful. Such radical styling, however, does draw attention to collections and serves a very important function with regard to publicity.

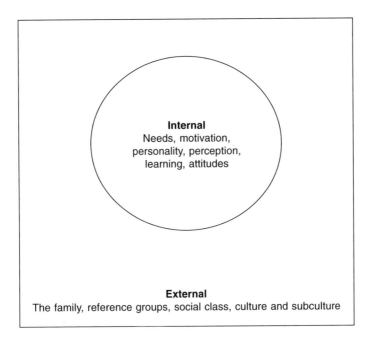

Fig. 8.3   Internal and external factors influencing product choice.

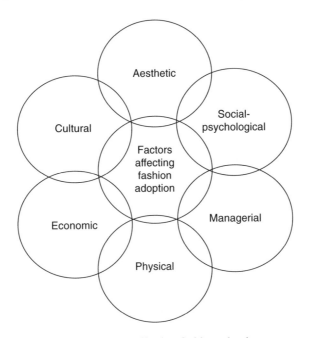

Fig. 8.4   Factors affecting fashion adoption.

Colours have different connotations and traditional uses across cultures, and this can have a huge impact for manufacturers and producers. For example, in Southeast Asia, where light blue symbolises death and mourning, the drink Pepsi-Cola lost its dominant market share to Coca-Cola when they changed the colour of their coolers and vending equipment from deep 'regal' blue to light 'ice' blue.

Just what are considered to be shameful parts of the body varies from culture to culture and generation to generation. In China, the female foot has been considered taboo, in Japan the back of the neck, in eighteenth-century France the offending body part was the shoulder, while the Indians of the Orinoco felt deep shame at the wearing of any clothes.

8.1.2.3   Social–psychological factors

Factors in this category include self and group identity, personal expression, reference groups, group expectations/peer pressure, role and social status. How people see themselves and how they wish to be perceived is reflected through the products they buy and the clothing they wear. Bankers wear sober dark grey business suits; they want to be seen as serious and reliable and grey suits are perceived as serious, unexciting clothing. If bankers were to come to work in bright coloured suits, they would be perceived as frivolous and it is unlikely they would be given business.

Often people will adopt the products used by, and the clothing worn by, those they admire – pop stars, film stars, sporting personalities. Queen Elizabeth I set a fashion for red hair while Princess Diana set fashions for spotty socks, woolly sweaters featuring sheep and even a hairstyle.

8.1.2.4   Economic and political factors

Factors that come into this category include technology, production and distribution of goods, consumer demand, income and price. Products have to be available, affordable and accessible. The fashion for drip-dry, non-iron clothing in the 60s came about only because of the technological developments of the time, notably the development of nylon.

8.1.2.5   Managerial factors

The factors covered in this grouping include needs and desires, management of resources, and buying habits and practices.

8.1.2.6   Physical factors

These factors include comfort, which is probably more of a minor consideration, certainly in clothing, than may at first be thought. Throughout history, many fashions have been quite impractical and most uncomfortable. Very tight jeans are uncomfortable and restrict movement, yet when fashion demands we all try to squeeze into jeans that are much too small for us. However, active sportswear must be comfortable and functional.

8.2   Fashion

Fashion is part of the social world we inhabit, we constantly make reference to it and are surrounded by shops that sell it. Fashion is news; it is considered frivolous and trivial yet it is big business. The fashion industry is concerned with style innovation, not just the production of adequate clothing. Fashion is more than a commodity, it is an attribute with which some styles are endowed. For a particular style to become a fashion, it has to be worn and recognised and acknowledged to be a fashion. This refers to clothing but there are fashions in everything — cars and shops, holidays and even pets.

Clothing has been said to serve four basic functions: to protect the body, to exalt the ego, to arouse emotions in others and to communicate by means of symbols. Other products equally serve these functions. The purchase of a brand new, top of the range sports car may have safety features that influence purchase decisions but more important factors are likely to be the pleasurable feelings that the purchase gives the buyer, mixed with the knowledge that others will be envious and that the driver will have a high status.

Stella Blum, Director of the Costume Institute of the Metropolitan Museum of Modern Art, has described a fashion as being 'no accident. It will always suit the occasion, time and place. Fashion is really a matter of evolution. It has a natural life.'

8.2.1   Why do fashions change?

Mystrom in *The Economics of Fashion* [1] thinks fashion change is triggered by boredom, and this is backed up by Hollander who in *Seeing Through Clothes* [2] suggests changes in fashion are motivated by visual need for something new. Other research, by Kroeber and Richardson [3], found two cycles of change: annual changes of a more trivial nature and longer cycles that were more subtle. They argued that social and political changes cause disruption of the style pattern. Table 8.1 shows the factors that impede and accelerate changes in fashion.

Table 8.1   Factors that impede and accelerate changes in fashion.

Impeding factors	Accelerating factors
Rigid class distinctions	An open class system
Poverty	Abundance and diffusion of wealth
Customs	Extended education
Isolation	Greater cultural contact
Fear of the new	The advance of technology
Government regulations	Improved status of women
Totalitarianism	Increased leisure

8.2.2   Forecasting fashion trends

'The word "fashion" is synonymous with the word "change". Fashion begins with fabrics and fabrics begin with colour. Each season colour must make a new statement for fashion to continue its natural evolution. Our eyes need refreshment. We may have disliked the way we looked in last season's clothes, but this season offers the opportunity to wear new colours, groom differently and appear as a new person. And if we are still dissatisfied, three months later there is a new season . . .'

This quotation is from Ed Newman of the American Textile Manufacturers' Institute and it can just as easily be applied to other textile products. If we do not like our surroundings, textiles can help us effect a transformation and if we do not like that, we can adopt the next trend to come along.

Consider the fabrics and fashions that are in the shops now for the current season. Fig. 8.5 shows that the designers who styled the garments would have been working on them 7–12 months ago, the fabric designers would have been working on their ranges 12–15 months ago, the yarn designers 18–20 months ago and the colourists approximately 24 months ago.

How did the designers know what people would want? What is 'in fashion' is constantly changing. How can designers predict what people will want? Where do they find directions and ideas for their new ranges?

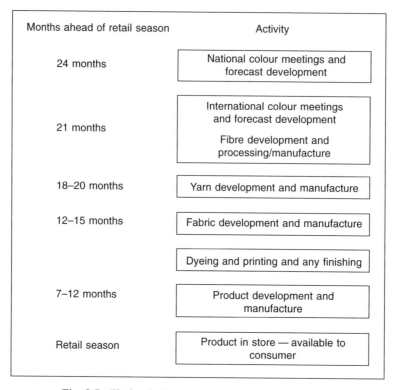

Months ahead of retail season	Activity
24 months	National colour meetings and forecast development
21 months	International colour meetings and forecast development Fibre development and processing/manufacture
18–20 months	Yarn development and manufacture
12–15 months	Fabric development and manufacture
	Dyeing and printing and any finishing
7–12 months	Product development and manufacture
Retail season	Product in store — available to consumer

Fig. 8.5   Timing in the textile and clothing industry.

The majority of designers work under quite considerable constraints. Certainly in the mass market they are often trying to produce designs which correspond to already identified trends. Ideas produced by designers undergo a ruthless process of selection by the designers themselves and within the producing company. The aim is to identify those designs most likely to be successful, that is those that correspond most closely to the anticipated fashion, as well as working within production and cost constraints.

The key for designers making trend decisions is to be as aware and as informed as possible. Knowledge of the market is important; those responsible must get out and about to see what is on offer. What is selling now? What has sold in the past? Can any trends be spotted from the data gathered; any conclusions drawn? Who are the competition, what are they doing and how successful are they? Magazines can help here.

Designers must have access to sales figures. What colours/fabrics/styles are selling well or have sold well in the past? The sales team should constantly feed information back to design. Who exactly is the customer? What is their lifestyle?

An understanding of the factors that influence fashion is essential to help recognise those that are likely to influence future fashion. The media — television, radio and the press — and theatre and cinema have a great influence on the buying public. In particular, television and the cinema have a great deal to do with the acceptance of new styles by consumers. James Dean, in the film 'Rebel Without a Cause', wore a white tee-shirt and denim jeans, setting a fashion that was to become a classic.

An awareness of what is happening in the world of cinema and television is also required. Who are the personalities in the news? In their time Elvis Presley, the Beatles and Jackie Kennedy have all had a tremendous influence on fashions; people love to emulate the personalities they admire.

What exhibitions or events are happening? What artists are in the news? Art collections on world tour often have an influence on fashions. When the Matisse and Chagall exhibitions were on tour and in the media spotlight, many areas of design reflected the styles of these painters.

What new technologies are available? The 60s fashion for easy-care fabrics could not have come about had the man-made fibres with the necessary properties not been developed.

Designers also influence fashion. What is happening in the world of design? Who are the leaders? Dior in 1947 gave the fashion world the 'New Look'; Courrèges in the 60s started the space-age look that became so popular, with his minis and white plastic boots; Vivienne Westwood gave the fashion world the mini-crini.

8.2.2.1   Colour and forecasting

Colour has come to dominate many industries. The textiles and clothing industry is certainly no exception. Research work carried out by yarn and textile manufacturers, fibre producers, retail groups and trade fair organisers consistently shows that the first response by a customer, whether textile buyer or retail shopper, is to colour.

Colour forecasting is the selection of ranges of colours that are predicted for a particular product/market at a particular time in the future. Many colour forecasts are specific to particular product ranges, i.e. men's knitwear, children's leisurewear, etc., but most will show three colour groupings — pales, mediums and darks. Within these groups there are likely to be several colours referred to as classics (colours which have been accepted over a long period of time, such as camel, navy, bottle green and black). As the vast majority of consumers do not replace their car or the products in their house and their clothing every season, colour ranges for a specific season must take into account the colours of the previous seasons as well as what might be described as new 'fashion' colours. Any colour palette will normally therefore show within its pales, mediums and darks, some of the previous season's fashion colours, the new season's fashion colours and some classics.

It has been demonstrated that there are many factors that affect consumers' colour choice. With increasing consumer awareness regarding the use of colour, it is critical that companies understand these factors and how they affect their particular market. Marketers need to understand the effect that colour has on consumers and colour forecasting in order to deliver appropriate colours for their particular market. Even a slight difference in shade from what is required by the final consumer can be catastrophic to the manufacturer.

8.2.2.2   Forecasting services

As well as information gathered by designers themselves, there are various organisations that offer advisory services.

One of the leading services for colour trend information is the International Colour Authority (ICA), who publish their forecasts twice a year, 24 months ahead of season. Other forecasting services include those by Infamoda Inc., Promostyle and Design Intelligence, who publish twice-yearly colour, fabric and styling ideas for all sectors of the industry. These come out approximately 12–16 months ahead of season.

The British Knitting and Clothing Export Council (BKCEC) hold twice-yearly seminars on colour and styling trends, approximately 14–16 months ahead of season. Organisations promoting fibres, both natural and man-made, provide promotional forecasting services for their respective fibres. The yarn spinners advise their customers, the fabric manufacturers, on the trends as they see them: the fabric manufacturers, in their turn, advise their customers, the garment manufacturers, and so on. Trends are edited and refined at each stage of the manufacturing process.

8.2.2.3   Trade shows and exhibitions

At every stage there are trade shows, where designers have the opportunity to see the forecast directions from a wide spectrum of sources.

The first shows are the yarn shows —

- Pitti Filati in Florence
- Expofil in Paris

then come the fabric shows —

- Interstoff in Frankfurt
- Première Vision in Paris
- Tissu Premier in Lille
- Fabrex in London

then the furnishing shows —

- Heimtex in Frankfurt
- Decorex in London

then the garment shows —

- ESMA in Milan
- Prêt-à-Porter in Paris
- Igedo in Dusseldorf
- The London Shows

It can be seen that many influences must be taken into consideration when predicting future trends; these include consumer moods, the political and economic climate, global trends, environmental issues and technology. Forecasters need to be aware of international influences and to constantly absorb the world around them. However, although consumer moods are considered, this is often in a global context and forecasters should also take into account aspects such as preferences for particular consumer groups.

8.3   Summary

As designers are designing for the future, they require an understanding of the factors which influence consumer choice and fashion; why people choose the products that they do, the factors influencing fashion adoption and the reasons why fashions change. Forecasting trends is an important element of textile designers' work. As well as gathering information for themselves through visiting trade fairs and exhibitions and keeping aware of what is selling now, designers can use the trend information and consultancy services provided by fashion forecasting organisations. Colour is an extremely important factor in product choice and designers need to understand the way colours are adopted. Selecting the wrong colours for a product range can have a disastrous effect on sales.

References

1 Mystrom, P.H., *The Economics of Fashion*, New York, Ronald Press, 1928.
2 Hollander, A., *Seeing Through Clothes*, New York, Viking, 1978.
3 Kroeber, A.L. and RICHARDSON, J., Three Centuries of Women's Dress Fashions; A Quantitative Analysis, *Anthropological Records*, 1940, **5,** 111.

Bibliography

Hann, M.A. and Jackson, K.C., Fashion: An Interdisciplinary Review, *Textile Progress*, 1987, **16**, 4.

Hollander, A., *Seeing Through Clothes*, New York, Viking, 1978.

Horn, M.J., *The Second Skin: an Interdisciplinary Study of Clothing*, 2nd ed., Boston; London, Houghton Mifflin, 1975.

Lurie, A., *The Language of Clothes*, rev. ed., London, Bloomsbury, 1992.

McDowell, C., *McDowell's Directory of Twentieth Century Fashion*, London, Muller, 1984.

Mystrom, P.H., *The Economics of Fashion*, New York, Ronald Press, 1928.

Perna, R., *Fashion Forecasting: Mystery or Method?*, New York, Fairchild Publications, 1987.

Rouse, E., *Understanding Fashion*, Oxford, BSP Professional, 1989.

Sproles, G.B., *Fashion; Consumer Behaviour Towards Dress*, Minneapolis, Burgess, 1979.

Sproles, G.B., Analysing Fashion Life Cycles – Principles and Perspectives, *Journal of Marketing*, 1981, **45**, 116.

9

Weave and woven textile design

In Chapter 2, the process of weave design was outlined. Here this process is considered in some more detail.

9.1 Weaving

Weaving is probably the oldest form of fabric construction. Known throughout the world, the process probably originated some 8000 years ago with Neolithic man. Woven fabrics consist of two sets of threads, the warp and weft, that are interlaced at right angles to each other (see Fig. 9.1). The warp threads run parallel to the selvedge down the length of the cloth, and each warp thread is known as an 'end'. The weft threads are called 'picks' and run across the cloth, working under and over the warp ends from selvedge to selvedge.

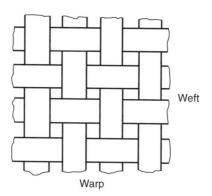

Fig. 9.1 Warp and weft.

Weaving involves three basic operations. First, some of the warp threads are lifted. This lifting of some threads and the leaving down of others creates a space called the shed. The weft yarns are then passed through this shed. Finally, the weft is beaten up to the fabric already formed. The loom is the device used for keeping the warp threads evenly spaced and under tension. Quite complex structures can be made on simple looms and essentially later developments have only made some of the weaving process easier and faster.

9.1.1   Initial considerations

The initial design briefing should give the designer the information needed to undertake the design work. From this, the designer should be clear about purpose, performance and price. All three of these will have an impact on the final designs:

- *Purpose.* The designer needs to know why the design work is being requested. What is it intended for — is it for a new range? If yes, for what season? Is the work to co-ordinate with existing styles? What is the intended customer profile?
- *Performance required.* How does the fabric have to perform? — e.g. if the fabric is for upholstery, what type of use is expected? Is it for occasional use or heavy use? The performance requirements for these two uses are clearly quite different.
- *Price.* A product going into a market where it can command a high price can use more expensive components than something that is intended as part of a cheaper, budget range.

9.1.2   Colour considerations

With their customer profile and the season they are designing for in mind, the weave designer will gather ideas for some initial design work. Early in the design process, it is likely that the weave designer will establish a likely range of colours. Ideas for colour will come from their own observations but it is likely that they will also use some of the colour trend information available in forecasting publications and on the web. As well as choosing colours to fit in with predicted trends, weave designers need to ensure that they choose colours that work together. For any company, holding large quantities of stock is expensive, tying up money and space, and the fewer the number of yarns, in terms of colour and qualities that can be used to achieve the desired effects, the better. Colour choice also has to bear in mind the requirements of existing and potential customers.

Colour palettes will often have four sections, lights/pastels, mid-tones, brights and darks, carefully chosen to satisfy the requirements outlined above.

9.1.3   Yarn selection

With colours selected, decisions will have to be made about yarns. Yarns may be sourced from ranges put together by spinners or they may be developed specially for the product being developed.

Many mills will weave a range of different fabrics although it is likely that some of these, at least, will continue through from one season to another. In the case of branded fabrics such as *Viyella*, the quality of yarn used will not change — any change in yarn range will only be in terms of colour, with some classic colours running from season to season and year to year and with some fashion shades being introduced on a regular basis to replace shades that have become less popular.

The type of yarn used in a fabric has a great impact on the final cloth. The fibres used for the yarn will affect handle and performance characteristics, as will the count (a measurement of linear density usually given as weight per measured unit of length or length per measured unit of weight) and the construction of the yarn itself. Fancy or effect yarns combined with different weaves and structures give an almost endless variety of possibilities for the designer, and with colour the possibilities are expanded even more. A variety of patterns, called 'colour and weave effects', may be achieved using coloured yarns in combination with various weaves.

9.1.3.1 Sourcing yarns

Mills will usually have regular suppliers of yarn although production departments will continually be looking out for new suppliers who can supply more cheaply regular yarns that run from season to season and of course new suppliers with exciting innovative ranges appropriate to their market. Yarn shows and trade fairs throughout the world form a showcase for yarns available and designers will normally visit such shows regularly to keep up to date with trends, see what is available and make contacts with new suppliers.

9.1.3.2 Considerations for yarns to be used in weaving

For weaving yarns, as well as trend and customer requirements having to be considered, the weaving process itself will impose some restrictions. By the very nature of the weaving process, warp yarns are under much more strain and tension than weft yarns. Breakages in a warp are expensive, slowing down production and requiring darning out after weaving. Warp yarns therefore are generally of a better quality and stronger than weft yarns which, supported by the warp as they are inserted, can be much weaker.

When yarns have to be over-dyed because the initial colour match is rejected, this over-dying will often weaken the yarn. While the weakened yarn may no longer be suitable for a warp, it can still be used in the weft.

In commercial woven fabric production many fabrics are woven with a different number of ends and picks per centimetre. Usually the number of warp ends per centimetre will be greater than the number of weft picks per centimetre. This is because although setting up the loom with a higher number of ends per centimetre will take longer, once the loom is set up, a heavier weft yarn will produce fabric more quickly than a finer yarn with more picks per centimetre.

Weave fabric designers need to be aware of the implications that their designs will have for production.

9.2   Weave structure

The interlacing pattern of the warp and weft is known as the weave. To create a weave, warp threads have to be lifted in the required order and weft yarns inserted in the space, or shed, that is created. The lifting of the appropriate threads is achieved by threading warp yarns in some sort of device that allows certain groups of threads to be lifted as necessary. In a simple loom the devices may be shafts, wooden frames with metal wires or healds, through which the warp yarns are threaded. Once threaded the appropriate shafts are lifted in the required order by some sort of mechanism.

9.2.1   Drafting and lifting plans

The draft is the instruction as to how the threads in the warp are drawn onto the shafts. Every thread that follows the same intersecting pattern throughout the length of the warp can be drawn onto the same shaft.

The lifting plan is the instruction for the lifting and lowering of the shafts. Different mechanisms have been developed to do this, including simple treadles and, for weaves that require more shafts, dobbies. In dobby shedding, the lifting of shafts is controlled by pegs in a series of strips of wood known as lags. These pegged lags, in conjunction with the dobby mechanism itself, cause the required shafts for the desired weave structure to be lifted. Lifting plans for dobby looms are therefore often referred to as 'peg-plans' and, in fact, this term is often used for lifting plans when mechanisms other than dobbies are used to lift the shafts. On jacquard looms, lifting plans were used to punch jacquard cards. However it is now more likely that jacquard looms, which have a greater pattern area and require a more complex patterning system than looms controlled by treadles or dobby mechanisms, will be controlled by some sort of CAD/CAM system. (CAM is 'computer aided manufacture'.)

9.2.1.1   Considerations in drafting

To help reduce problems in weaving, particularly with broken ends, heavy shafts (that is those that have many warp threads drawn through them) should be kept to the front. Weak yarns and tight working ends, wherever possible, should also be drawn on the front shafts. This is because the shed created when the front shafts are lifted will be larger than the shed created when the rear shafts are lifted.

The draft should be kept as simple as possible so that it is easy to follow, and an appropriate number of shafts should be used: too many healds per centimetre causes friction and therefore breakages, and this is the reason why for plain weave four shafts are more commonly used than two.

9.2.2   Denting

The reed is a device through which the warp yarns are threaded in order to keep the threads spaced correctly during weaving. The dents are the spaces in the reed through which the yarns are threaded in groups of two, three or more depending on the number of ends per centimetre required, the type of yarn being used and the weave.

Denting plays an important part in achieving a flawless fabric. Too many threads in one dent can cause breakages and a reed spacing should be selected that is wide enough to allow a knot in the yarn to slip through easily.

The number of dents per centimetre of the reed being used, and the way the ends are dented, determines the number of ends per centimetre.

9.2.3   Sett

Sett is the term used to describe the spacing of the weft and warp threads in a woven fabric and is usually expressed as numbers of threads per centimetre.

If the number of weft threads, or picks, per centimetre is the same as the number of warp threads per centimetre, then the fabric is said to be 'square sett'. The density of threads and the relationship between the ends and picks per centimetre affects the type of fabric produced. Altering the sett of a fabric changes the appearance and the feel, or handle, of the fabric. 'Scrim' is a very loosely woven fabric with the warp and weft threads spaced out.

9.2.4   Weave repeat

One weave repeat contains the smallest number of different intersections between warp and weft that, when repeated in either direction, gives the weave of the whole fabric.

9.2.5 Warping and picking plans
These are the orders in which warp and weft yarns are arranged.

9.2.5.1 Warping plan
The warping plan is the order in which the warp threads are arranged. The smallest number of ends in colour and/or count that repeats across the fabric is the warp repeat. The warping order is given from left to right, standing at the front of the loom.

9.2.5.2 Weft or picking plan
The weft or picking plan is the order the weft threads are inserted into the warp. The smallest number of picks in colour and/or count that repeats up the fabric is the weft repeat. The picking order is given from bottom to top, standing at the front of the loom.

9.2.6 Notation systems for weaves
To describe weaves is difficult and so a notation system is used which uses crosses or marks to indicate where the warp thread is uppermost. This graphical representation of interlacings is marked on point paper, a special graph paper for designing woven fabrics. The standard point paper used is ruled in groups of 8×8, separated by thicker bar lines. Each vertical space represents a warp end, and each horizontal space a weft pick. Each square therefore indicates an intersection point of one end and one pick. Only warp floats or lifts are indicated by a mark: a blank square represents a weft float.

9.3 Plain weave

Plain or 'tabby' weave is the simplest and most frequently used weave there is. In this, the threads in both warp and weft directions interlace alternately (see Fig. 9.2). Because it has the maximum number of interlacing or binding points possible, it is firmer and stronger than a fabric that is otherwise identical but made in another weave structure.

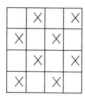

Fig. 9.2 Notation of plain weave.

Lightweight fabrics with a loose setting have to be made in plain weave otherwise the ends and picks would slip over each other and cause the fabric to distort. Chiffon and voile are two lightweight fabrics made using plain weave. Voiles are normally produced from cotton or linen, while chiffon is traditionally made from silk.

Plain weave is very versatile and a variety of decorative effects can be achieved with plain weave structures. The introduction of coloured yarn can create checks and stripes, and other colour and weave effects. The introduction of yarns of varying profile and construction will also add interest and, by cramming more ends in some dents in the reed than others, raised or crammed stripe effects can be created.

Altering sett also creates different effects. The effect of ribs can be given to plain weave fabrics by having more ends than picks per centimetre, or vice versa. More ends than picks will give ribs running in the weft direction. The warp will be more visible than the weft and so the fabric is described as warp-faced with a weft-ways rib. Comparing the threads taken from both directions will show the ends to be more heavily crimped than the picks. The ends do more bending round while the picks lie across the cloth. Poplin, taffeta, and grosgrain are all plain weave fabrics that are warp-faced with a weft-ways rib. Poplins are normally made from cotton or linen, while taffetas and grosgrains are traditionally made from silk.

Tapestry weave is a weft-faced plain weave where the weft threads are packed closely together so the warp is hardly seen. Tapestry is often associated with the large pictorial wall hangings of medieval and later European origin however, strictly speaking, tapestry is simply a distinctive woven structure — a weft-faced plain weave with discontinuous wefts.

9.3.1   Plain weave colour and weave effects

Introducing colour patterns in the warp and weft of plain weave gives rise to different patterns or colour and weave effects.

In the plain weave effects shown in Figures 9.3–9.5, where a cross is indicated the warp will be uppermost; therefore, where there is a black warp thread indicated, every cross in that vertical column will show as a black thread uppermost. A square that is blank indicates a weft thread uppermost; and so where a black warp thread is indicated, in that horizontal row every blank will show as a black weft thread being uppermost.

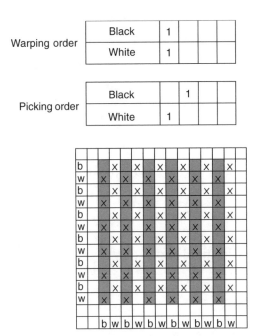

Fig. 9.3   Colour and weave effects — warp hairline.

Warping and picking order

Black	1			
White	1			

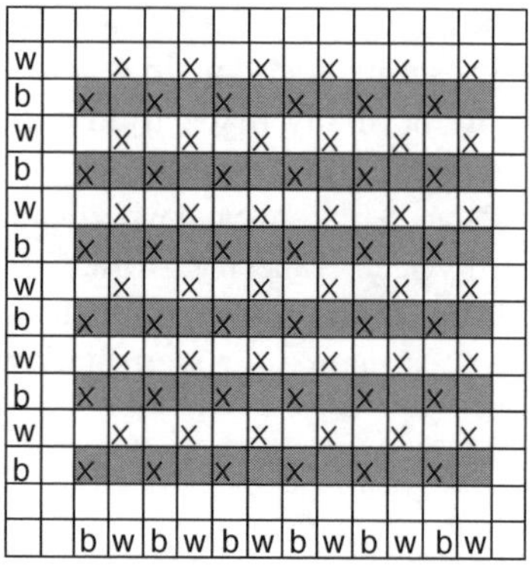

Fig. 9.4　Colour and weave effects — weft hairline.

Warping and picking order

Black	2			
White	2			

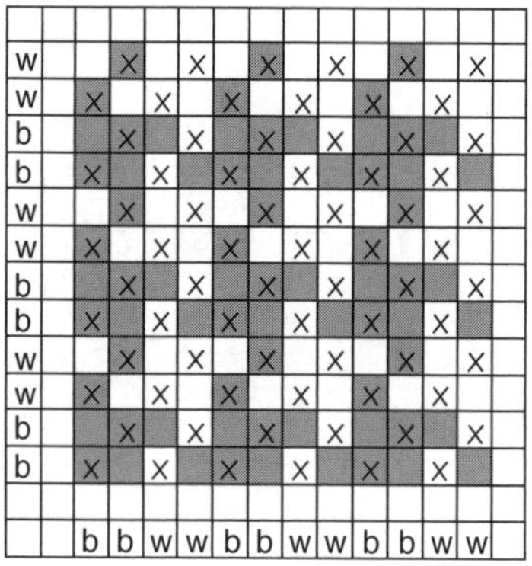

Fig. 9.5　Colour and weave effects — four-point star.

9.4　Some simple basic weaves

There are many other weaves that are commonly used by designers and some of the most basic of these are described next.

9.4.1　Hopsack or matt weaves

Hopsack or matt weaves are based on plain weave but with two or more threads working in the same order in both warp and weft. The fabrics have a smooth surface with a greatly enlarged plain weave structure. If two fabrics of otherwise similar construction are woven as plain weave and 2/2 hopsack, the hopsack will be less stiff because of its fewer intersections. To prevent adjacent threads rolling over each other during weaving, it is usual to have only two threads through each dent, with those in the same dent working in different orders in the warp. In Fig. 9.6 the first thread would be drawn through a space in the reed, the second and third threads would be threaded through the next space etc.

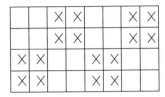

Fig. 9.6   Notation for 2 × 2 hopsack.

9.4.2   Twill weaves

Twill weaves are characterised by continuous diagonal lines and can be more closely sett than their equivalents in other weaves. More weight and better drape can be achieved with twills. Twill weaves are less stiff than equivalent plain weave fabrics with the same yarn and sett.

In 'Z' twill, the diagonal runs up to the right; in 'S' twill, the diagonal runs up to the left (see Figures 9.7 and 9.8). The smallest number of threads on which it is possible to construct a twill is three. Figure 9.9 shows a 2 × 2 twill right.

Fig. 9.7   'Z' twill.

Fig. 9.8   'S' twill.

Fig. 9.9   Notation for 2 × 2 twill.

9.4.3 Colour and weave effects on twills
Just as colour and weave effects can be created on plain weave, these can be created on twill weaves.

9.4.4 Pointed and herringbone twills
Pointed and herringbone twills are produced by combining twills to the left and twills to the right. The joins between the twills in herringbones are said to be 'clean cut'—the warp threads on either side of the join work in opposition. (See Figures 9.10 and 9.11.)

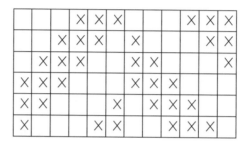

Fig. 9.10   3/3 herringbone 6R6L clean cut.

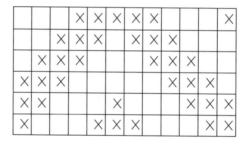

Fig. 9.11   3/3 pointed twill 6R6L.

9.4.5 Regular satin and sateen
Typically fabrics made from these weaves have a smooth and lustrous appearance. A satin is warp-faced, while a sateen is weft-faced. The intersections of warp with weft are distributed so that they are never adjacent. In a closely-woven cloth, the floats conceal the binding points. Five- and eight-end satin and sateen weaves are common.

The smallest number of ends and picks on which it is possible to construct a satin/sateen is five. Satinettes are created on four ends and four picks; these are not true satins/sateens as there are adjacent binding points. (See Figures 9.12 and 9.13.)

Fig. 9.12   Weft satinette.

Fig. 9.13   Warp satinette.

Warp satins have a substantially higher number of ends than picks. The warp yarn very often consists of fine silk or worsted while the weft may be an inferior quality yarn. Weft sateens have a higher number of picks than ends. The weft is of a superior quality and covers the warp. Production costs are high because of the high number of picks. Satins and sateens are frequently combined in jacquard weave designs to create patterns.

9.5   More complex weaves and weave combinations

The possibilities for designers of woven fabrics in terms of weave structure are almost endless; they can create their own weaves and there are a host of more complex structures they can use and adapt as they see fit. However, it is worth noting that, as with so much design, the simplest ideas are often the best. This is why the vast majority of woven designs use simple structures, often combined with plain weave.

9.6   Sample warps

Because of the length of time it takes to set up a warp, much initial weave design work is done on section warps (see Chapter 2) with colours chosen and warp threaded for flexibility. Often, fabrics that will be woven in production as warp stripes may be sampled on a solid warp, with the stripe being introduced in the weft. Using a solid warp in this way allows many different striping ideas to be tried in the weft. Introducing stripes in the warp of a sample blanket limits the sampling that can be done since that stripe will feature through the whole of the warp unless tied out. In production, stripes are often produced running in a warp-ways direction. This is because once the loom is set up, a plain weft is faster to weave than a patterned weft.

9.7   Finishing

After fabrics have been woven, they may be processed (finished) to enhance appearance and/or performance. Wool fabrics are normally treated or scoured after weaving to remove natural fats and waxes as well as dirt, oil and other impurities. A high lustre may be achieved by calendering, a process where the fabric is passed through two heavy rollers on a machine known as a calender. Fabrics may be treated with chemicals to improve performance such as resistance to soiling or to fire.

9.8   Fabric specifications/making particulars

For all designers it is important that they record exactly what they have done — the yarns they used, the draft and lifting plan, etc. (see Chapter 5). This is to enable the production office to reproduce initial samples as sample lengths and as mass-produced fabric, and to allow fabrics to be costed.

9.9   Summary

Woven fabrics are made by interlacing two sets of threads set at right angles. The threads running in a vertical direction and held under tension on the loom are known as the warp, while the threads introduced across the warp are known as the weft. The resultant characteristics of any woven fabric (the look, the handle and how it performs) come from a combination of several elements; the yarns and fibres used, the way these are arranged in terms of colour and sett, the way the threads are interlaced (the weave) and the finishing processes applied to the fabric after weaving.

Weave design often starts with colour — colours chosen by the designer from their own feelings as to what would be right, and from colour forecasts. With colours selected, the weave designer sources yarns, taking into account several factors including considerations in drafting. After some initial work on paper and/or on a CAD system, sample warps will be produced and blankets woven. Using colour with different weaves gives rise to a variety of different colour and weave effects. There are a multitude of different weave structures that can be tried out.

Plain weave is the simplest and most widely used of all weaves. With the most interlacings, or binding points, it is very strong and is used for a variety of different fabrics, from rough sackings and other industrial textiles through to light chiffons and sheer voiles. Changing the relationships between ends per centimetre and picks per centimetre, and introducing colour patterns in the warp and weft, create different fabric effects, as does applying different finishes. Full making particulars or fabric specifications have to be recorded by the designer to allow their samples to be reproduced.

Bibliography

Goerner, D., *Woven Structure and Design — Part 1: Single Cloth Construction*, Leeds, British Textile Technology Group, 1986.

Goerner, D., *Woven Structure and Design — Part 2: Compound Structures*, Leeds, British Textile Technology Group, 1989.

Harris, J., *5000 Years of Textiles*, London, British Museum Press, 1993.

Hatch, K.L., *Textile Science*, New York, West Publishing Company, 1993.

Paine, M., *Textile Classics: a Complete Guide to Furnishing Fabrics and their Uses*, London, Mitchell Beazley, 1990.

Watson, W., *Watson's Textile Design and Colour: Elementary Weaves and Figured Fabrics*, 7th ed., London, Newnes-Butterworths, 1975.

10

Weft knitting, weft-knitted fabric and knitwear design

In Chapter 2, an overview of the process of knit fabric and knitwear design was outlined. This overview is expanded here.

10.1 Knitting

In knitting, the fabric is formed by intermeshing loops of a single yarn or set of yarns together. The origins of knitting are rather obscure but some of the earliest examples of true knitting are stockings from Egypt which are dated at around 1200–1500BC. However, it was not until 1589 that the process was mechanised, when the Reverend William Lee of Calverton in Nottingham in the UK was credited with the invention of the knitting frame. Since a single thread was used and the formation of loops took place in a widthways direction, the process was known as weft knitting (see Fig. 10.1). In the 1760s Crane, again of Nottingham, modified William Lee's frame, giving each needle a separate thread. The threads lay in a lengthways direction so the process was known as warp knitting (see Fig. 10.2).

Fig. 10.1   Weft knit structure.

Fig. 10.2   Warp knit structure.

When knitting is done by hand, the loops are held on one knitting needle and the yarn is pulled through each loop individually to form the new row of stitches. This row of stitches is subsequently connected to the next row. In machine knitting there is a needle for every stitch and it is the action of this needle that pulls the yarn through the old loop to form the new stitch.

As with woven fabrics, there are several elements in any knitted fabric that together give the resultant fabric its characteristics in terms of appearance, feel and performance. The yarns used and the fibres that these yarns have been made from will affect the final fabric, as will the yarn colour and the order in which the yarns have been arranged. The way the threads, or yarns, interlace (the knit fabric structure) and the machine gauge (the size and number of needles per inch) will also have an impact on the ultimate fabric, as will any processes that the fabric undergoes after knitting.

10.2   Weft-knit manufacture

As with other textile products, certain countries and geographical regions within countries have become home to the knitting and knitwear sectors of the textile industry. In the UK, the weft-knitting and knitwear sector of the industry is located mainly in the Midlands and the Scottish Borders region. Scotland is where much of the production of the top-quality fully-fashioned (shaped on the machine) pure wool and cashmere knitwear is found, manufactured by companies such as *Pringle*, *Ballantyne*, *Barrie* and *Braemar*. The Midlands of England, particularly Leicester, is where many of the mass market producers that supply to the high street stores are found. Traditionally, the manufacturers in this area mainly make cut and sewn knitwear, that is, garments made from body and sleeve blanks (garment panels) that are cut to shape and then sewn together. Other companies, again mainly in the Midlands in the UK, make knitted fabric, called jersey, which is produced in wide widths from continuous-fabric circular machines. Sweatshirt fleece and other sports fabrics can be classed as jersey fabrics.

There are also several companies whose business is the merchanting of knitwear. Such companies do not actually manufacture knitwear themselves; instead they will buy their production from a variety of manufacturers throughout the world. Able to select their manufacturing plant according to market demands, these companies can offer a wide spectrum of knitwear from heavy handknits to firm lambswools.

10.3 Machine gauge

Knitting machines can be grouped in many ways. One way is by the gauge—the number of needles per inch or centimetre. The exception to this is for fully-fashioned machines, where the gauge is often expressed as the number of needles per one and a half inches. A heavy-gauge fabric is one that has been knitted on a machine with a few needles per inch ($2^1/_2$ to 5 needles per inch are common gauges in the UK). A fine-gauge fabric is one that has been knitted in fine yarn on an 18- to 32-gauge machine (UK).

'Knitwear' (knitted garments usually with integral ribs) is normally produced on $2^1/_2$- to 14-gauge machines, with traditional type knitwear (lambswool, botany and cashmere) more likely to be in the 7- to 12-gauge range. Jersey (continuous fabric) is produced on finer gauge machines, from 18–32 gauge.

10.4 Weft-knitting machines and fabric types

Weft-knitted fabric can be produced on either circular or flat machines. Circular knitting machines have the needles arranged in slots around the circumference of a cylinder and the fabric is manufactured as a tube. Flat machines have the needles arranged in a straight line on a needle bed with knitting taking place from side to side, the resulting fabric being produced in open width.

On a simple flat machine, a carriage carrying the yarn supply traverses back and forwards across the width of the machine. For each traverse, one row, or course, is normally knitted. The productivity of this type of machine is limited by the speed of traverse and the need to slow down and stop at either end. On circular machines the yarn always travels in the same direction and so the productivity is increased. Circular machines also allow an increase in the number of feeder points around the machine, with productivity being increased in direct proportion to the number of feeders. It is not uncommon to have 100 feeders around a modern 30-inch diameter machine.

Stripe effects can be achieved by feeding different-coloured yarn to different feeders. As well as being able to produce stripes, pattern and texture can be introduced using fancy yarns and modified loop shapes.

Fully-fashioned machines are flat machines that knit to shape individual garment pieces. A typical fully-fashioned machine will produce three individual garment backs or fronts across the width of the machine.

Flat machines usually produce body blanks, i.e. panels with integral ribs that are then cut and sewn into garments. Some circular machines can also produce body blanks. These panels are pressed on frames before slitting and cutting into body and sleeve pieces. Many circular machines however produce continuous fabric which is then slit and treated in much the same way as woven fabric. An exception to this is sweatshirt fleece and single jersey for tee-shirts, which is often sold in tubular form. Circular machines producing continuous fabric are usually large-diameter machines with many yarn feeds that knit at high speed.

Knitwear is traditionally produced on coarse to medium gauge fully-fashioned, circular or flat machines which can be programmed to knit a welt (a neat secure starting edge), rib and body fabric in sequence. Jersey fabrics are produced as continuous fabrics on fine-gauge circular machines. These fabrics can be divided into single or double jersey depending on whether the fabric is knitted on one or two sets of needles.

Over the last twenty years, developments in knitting have been considerable. In many instances these developments have been directly reflected in the trends seen in knitwear design, as designers have worked to produce new looks and styles. The introduction of machines with individual needle selection in the late 70s saw many garments being produced in designs reflecting this technology, e.g. large picture patterns of racing cars and animals on the front and back of garments. The latest commercially-available electronically-controlled flat machines with sophisticated needle control now enable complete garments to be knitted in one piece.

10.4.1   Plain fabric

The simplest weft-knitted fabric is made on one set of needles with all the loops intermeshed in the same direction. It is called plain fabric. The face of the fabric is smooth and shows the side limbs of the loops as a series of interlocking 'v's (see Fig. 10.3). The reverse is rough and looks like columns of interlacing semicircles (see Fig. 10.4). Plain fabric can be unroved (unravelled) from either end. It has a tendency to curl towards the back at the sides and towards the front at the top and bottom.

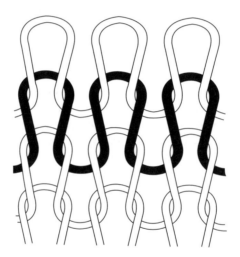

Fig. 10.3   Loop diagram showing face of plain weft-knit fabric.

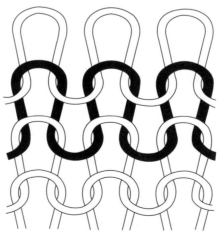

Fig. 10.4   Loop diagram showing reverse of plain weft-knit fabric.

10.4.2   Rib fabrics

Rib fabrics are knitted on machines with two sets of needles. These needles are arranged in such a way as to allow them to intermesh when raised, and this needle arrangement is called rib gaiting. Flat machines with two sets of needles arranged in this way are usually called 'v' beds because from the side they look like an inverted 'v'. The needle beds are called the front and the back beds.

Circular machines with two sets of needles have a dial and cylinder. The cylinder needles are arranged vertically round the machine and are the equivalent of the flat machine's front bed. The dial needles are arranged horizontally inside the cylinder and are the equivalent of the flat machine's back bed.

The simplest rib fabric is 1 × 1 rib. This has a vertical rib appearance because the face loop wales tend to move over and in front of the reverse loop wales. 1 × 1 rib has the appearance of the technical face of plain fabric on both sides until stretched, when the reverse loop wales in between are revealed. (See Fig. 10.5.)

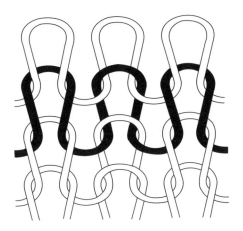

Fig. 10.5   Loop diagram: 1 × 1 rib.

Because 1 × 1 rib is balanced by alternate wales of face loops on each side, it lies flat, without curl, when cut. It can only be unroved from the end knitted last. This type of rib is an elastic structure with good widthways recovery after it has been stretched because the face loop wales move over and in front of the reverse loop wales.

Because ribs cannot be unroved from the end knitted first and because of their elasticity, they are particularly suited to the edges of garments such as the tops of socks, cuffs and the waist edge of garments.

10.4.3   Purl fabrics

Purl fabrics are knitted on machines with one set of needles, which are double-ended, allowing loops to be intermeshed in two directions. Purl fabrics are characterised by the fact that they have face and reverse loops in the same wale. This type of structure can only be achieved on purl machines or by rib loop transfer. Rib machines will knit purl structures if loop transfer between the beds is possible. Loops on the front bed can be transferred to needles on the back bed and *vice versa* to produce face and reverse loops in the same wale.

On a purl machine, the tricks (the slots in which the needles are located) of the two needle beds are directly opposite and in the same plane. This allows the double-ended needles to be transferred across from one needle bed to the other, enabling fabrics to be made that have face and reverse loops in the same wale (see Fig. 10.6).

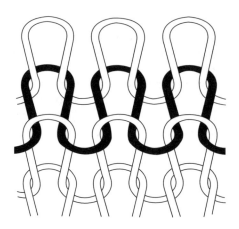

Fig. 10.6   Loop diagram: 1 × 1 purl.

There are two types of purl machine—flat purls, which have two horizontally opposed needle beds; and circular purls, which have two superimposed cylinders one above the other so that the needles move in a vertical direction.

10.4.4   Interlock fabric

Interlock machines are weft-knitting machines that have two sets of needles on a back and front bed (or dial and cylinder). Unlike rib machines however the needles in one bed are directly opposite those on the other bed. The fabrics are constructed so that opposite needles are not lifted at the same time. The fabrics produced are double-knit structures; essentially two fabrics interlocked together (see Fig. 10.7).

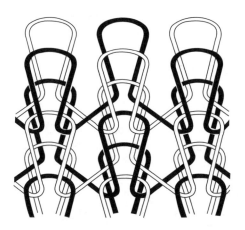

Fig. 10.7   Loop diagram: simple interlock.

10.5 Characteristics of weft-knitted fabrics

Knitted fabric is unique in that it possesses a high order of elasticity and recovery. Unlike woven fabric, which possesses a low degree of elongation, knitted fabric can be stretched to a considerable length and yet, when it is released, it will gradually return to its original shape and configuration. It is this feature of the fabric, plus the air permeability arising from its looped structure, that imparts to it the following properties: a high degree of wrinkle resistance (knitted apparel generally requires little ironing); good drape; a high degree of comfort; a porous nature allowing the skin to breathe freely; and elasticity allowing freedom of movement.

10.6 Weft-knitted fabric structures

There are several types of knitted stitches that are used to make weft-knitted fabrics. These stitches, in conjunction with the type of machine used, create a wide variety of knit structures for the designer to use and adapt.

10.6.1 Miss stitch

It can be arranged that a needle is not raised to receive new yarn. This results in a float at the back of the fabric when knitted on a machine with one set of needles (see Fig. 10.8). This miss or float stitch is a convenient way of hiding coloured yarns at the back of the fabric when these are not required on the face.

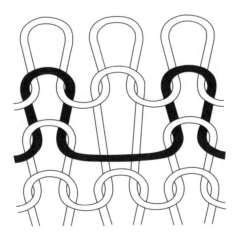

Fig. 10.8 Loop diagram: miss stitch.

10.6.2 Tuck stitch

Tuck stitches are formed when the yarn is taken into the needle but not fully formed (see Fig. 10.9). Tuck stitches offer many decorative possibilities; they can be used to produce openwork effects, make surface texture effects and improve ladder resistance.

10.6.3 Transfer stitch

The sideways transfer of loops over a number of courses provides the starting point for

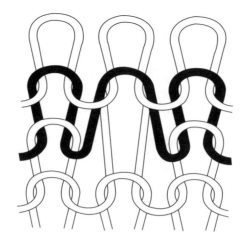

Fig. 10.9   Loop diagram: tuck stitch.

a whole range of structures (see Fig. 10.10). The inclined loop can be used for decorative effect and, if the spaces that the transferred loop would have occupied are left free, a wide range of openwork, lace and eyelet fabrics can be developed.

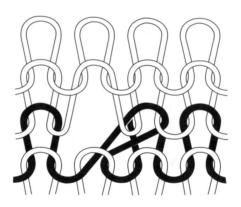

Fig. 10.10   Loop diagram: transfer stitch.

If the transfer operation takes place inwards from the edge of the fabric, the fabric can be narrowed, a few wales at a time, giving the possibility of producing a shaped or fashioned garment.

10.7   The graphic representation of fabrics

10.7.1   The face loop stitch
On a plain single-bed fabric, one side comprises face loops. This side shows the side limbs of the needle loops as a series of intermeshing 'v's. This stitch is formed on the front bed or cylinder and is graphically represented in Fig. 10.11.

Fig. 10.11   Face loop stitch.

10.7.2   The reverse loop stitch

This is how the stitch appears on plain fabric on the opposite side to the face loop side. The stitch is formed on the back bed or dial (see Fig. 10.12).

Fig. 10.12   Reverse loop stitch.

10.7.3   Notating knitted fabrics

The description of fabrics is often very lengthy and difficult to understand. Producing graphic representations in the form of loop diagrams is very tedious and so different ways of notating fabrics have been developed. The technical face of a plain fabric can be notated in two ways, as shown in Fig. 10.13, and the technical reverse as in Fig. 10.14.

Fig. 10.13   Face stitch notations.

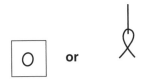

Fig. 10.14   Reverse stitch notations.

10.7.4   2 × 2 rib

On a rib machine, the needles are usually set out two needles in action and one out of action, with the two in action on one bed being opposite the one out of action on the other bed, as in Fig. 10.15.

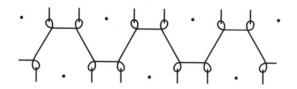

Fig. 10.15   2 × 2 rib on a rib machine.

This structure is also called Swiss rib. In some areas it is called 2/3 rib because two out of three needles in each bed are knitting. Sometimes it is called 2/1 rib because the layout of needles along each needle bed is two needles in action and one needle out of action. 2 × 2 rib produced on a purl machine is called English rib (Fig. 10.16).

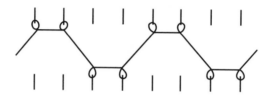

Fig. 10.16   2 × 2 rib on a purl machine.

10.7.5   1 × 1 purl

The simplest purl structure is 1 × 1 purl, which is the garter stitch of hand knitters. This purl consists of alternate courses of face and reverse loops.

Both sides of the fabric are similar in appearance. The lateral stretch is equal to plain knit fabric but the lengthways elasticity is almost double. When relaxed the face loop courses cover the reverse loop courses, making it much thicker than plain fabric. 1 × 1 purl can be unroved from both ends because the free sinker loops can be pulled through the bottom of the fabric. (See Fig. 10.6.)

10.7.6   Moss stitch

For moss stitch, in each course (row) the needles knit alternately face and reverse loops (see Fig. 10.17).

Both sides of the fabric are similar in appearance, and the widthways and lengthways stretch are both greater than that of plain fabric. It is thicker and bulkier than an equivalent plain fabric and it can be unroved only from the end knitted last.

10.8   Knitwear production

There are two main categories of knitwear production — fully-fashioned and cut and sewn.

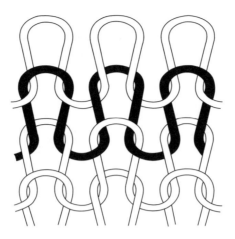

Fig. 10.17   Loop diagram: moss stitch.

10.8.1   Fully-fashioned knitwear

In fully-fashioned knitwear production, the garments are made up from pieces that are knitted to the correct size (fashioned) and shape required for the garment, while in cut and sewn knitwear, garments are made from pieces cut from panels, or lengths of fabric. In fully-fashioned garment construction, the garment pieces are joined together by a sewing process called overlocking. The overlock machine has a knife, which cuts off the edge of the fabric; stitches are formed round the raw edge preventing the fabric from unravelling. Fully-fashioned garments will usually be cup-seamed together and the neck and other trimmings will normally be attached by a technique called linking. Linking involves each individual loop on a rib being put on a point before being stitched to the body.

Designers of fully fashioned knitwear need to specify the exact number of needles required at any point in a garment panel so that panels can be knitted to the correct shape required.

10.8.2   Cut and sewn knitwear

Most cut and sewn knitwear is made from pieces of fabric that have integral ribs. The machines used to produce the panels are able to be programmed to knit a welt, a rib section and then the body fabric. Some machines also have the facility to knit patterns in the body fabric. Pieces of fabric with integral ribs are usually referred to as 'blanks'. Body blanks are the pieces from which the front and back garment sections are cut, while sleeve blanks are those from which the sleeves are cut.

Garment blanks can be produced on both flat and circular machines. Flat machines produce open-width panels and these will be knitted to the correct width, for example, for three body pieces and three sleeve pieces. The number of needles necessary for the bodies and sleeves is usually specified and needles are taken out of action at the appropriate places to give a guide to the cutters.

Circular machines produce tubes of fabric. The diameter of the machine, the gauge and type of structure knitted are some of the factors that determine the width of tube obtained.

Body and sleeve blanks will have an area where the pattern steps and this is where the blank is slit at cutting. Depending on the size of the blank and the pattern width, it is usual to get 2 or 3 body pieces round and 3 to 4 sleeve pieces. As with open width-panels from flat machines, it is usual to take needles out of action where a guideline is required to let the cutter know where to slit the panel. A guideline will come at the halfway point if a back and front can be cut from one blank. Depending on how many body or sleeve panels can be cut from a tube, the guidelines will split the panel into thirds or quarters.

The body and sleeve blanks come from a knitting machine joined together by drawthreads. These threads are pulled out, splitting the panels. As the knitted panels are knitted under tension, they have a tendency to relax when removed from the knitting machine. To complete this relaxation process the fabric is usually steamed. With fabric made from acrylic yarn, it can be seen to relax when steamed and this improves the handle. Blanks that are knitted in 100% wool, or with a high percentage of wool, need to be scoured to achieve a relaxed state with a satisfactory handle.

Garments are usually processed in batches; in the UK these batches usually consist of 12 garments or the components for 12 garments. One garment is often then referred to as 1/12, two garments as 2/12 and three as 3/12 etc.

When an order is taken the production office will allocate yarn and issue production tickets. The first place to process these tickets will be the yarn store, where the yarn will be issued to knitting with the appropriate tickets. The knitters will knit the required order, batching it up in dozens, i.e. enough body and sleeve blanks with the appropriate trimmings to make one dozen garments.

Different types of knitted trim may be required, depending on the style of garment required. Neck trims include crew necks, cowls, vee necks, shawl collars and collars and plackets. These may be folded or single and are often in a rib construction such as 2×2 or 1×1. For cardigans the trims can either be stolling or strapping. Stolling is the name given to rib trim put on so that the rib is at right angles to the direction of the rib on the body fabric. Strapping is narrow width rib trim, usually 1×1 rib or of a half milano construction.

After knitting, the blanks will be pressed and moved through to garment making-up where they are cut to the required shapes and made up into garments via a variety of sewing processes. The finished garments are pressed and checked before being folded up and bagged, ready for distribution.

10.9   Summary

In knitting, the fabric is created by intermeshing loops of yarn together. When a single yarn is used and the loops are formed in a widthways direction, the process is known as weft-knitting; when a set of warp threads running in a lengthways direction are intermeshed with all the loops in a row being formed simultaneously, this process is known as warp-knitting. As with woven fabrics, there are several elements that give a knitted fabric its ultimate appearance and feel, and impact upon its performance.

The different types of knitwear and knitted fabric manufacture have developed in different geographical areas. In the UK, the Scottish Borders is home to several manufacturers of fully-fashioned and intarsia knitwear using luxury fibres; in the Midlands, traditionally the manufacturers of cut and sewn knitwear for the mass market are to be found, as are manufacturers of jersey fabrics.

Knitting machines and the fabrics produced can be grouped in many different ways. These include the number of needles per unit length, whether the weft or warp-knitting process is used, what type of machine has been used (whether fully-fashioned, circular or flat) and how the needles in the machine are arranged.

The simplest weft-knitted fabric is called plain fabric; it is knitted on one set of needles and all the loops intermesh in the same direction. Rib fabrics are made on machines with two sets of needles, arranged so as to allow these to intermesh when raised. Purl fabrics are made on machines with double-ended needles that allow fabrics to be produced with face and reverse loops in the same wale. Interlock fabrics are made so that the knitted loops intermesh in such a way as to make the fabrics less extensible and more difficult to unrove. Knitted fabrics possess a high degree of stretch and recovery, a high degree of wrinkle resistance, generally drape well and are comfortable to wear.

In weft-knitting there are a number of basic stitches that, combined together on the different types of machines available, make an endless variety of structures. The main stitch structures used, in addition to face and reverse loops, are the miss stitch, the float stitch, the transfer stitch and the spread stitch. There are two main notation systems used to represent weft-knitted fabric structures.

There are two main categories of knitwear production; fully-fashioned and cut and sewn. In fully-fashioned knitwear production, the garment pieces are shaped on the machines, while in cut and sewn knitwear production the garment pieces are cut to shape from the fabric produced.

Bibliography

Harris, J., *5000 Years of Textiles*, London, British Museum Press, 1993.

Hatch, K.L., *Textile Science*, New York, West Publishing Company, 1993.

McIntyre, J.E. and Daniels, P. N. (eds), *Textile Terms and Definitions*, 10th ed., Manchester, Textile Institute, 1995.

Smirfitt, J.A., *An Introduction to Weft Knitting*, Watford, Merrow, 1975.

Spencer, D.J., *Knitting Technology*, 3rd ed., Cambridge, Woodhead Publishing, 2001.

Taylor, M.A., *Technology of Textile Properties: an Introduction*, 3rd ed., London, Forbes Publications, 1990.

11

Printing and printed textile design

In Chapter 2, the process of print design was outlined. This chapter looks at the way fabrics can be coloured, and some of the main types of print designs and layouts that can be used by print designers.

11.1 Printed textiles

Printing is rather an ambiguous term. Methods of colouring some areas of fabrics differently to others by using dyes, pigments and paints all tend to be termed as printing although, in fact, some of these are not strictly printing but rather dyeing and colouring techniques.

The appearance of a printed textile is affected by several different elements; the base fabric on which the pattern is made, the design and the way the unit repeat of this is repeated across the fabric, the types of dyestuffs applied and the way these are applied. A basic introduction to woven and knitted fabric construction is given in the previous two chapters and there is an introduction to simple repeating patterns in Chapter 4. This chapter looks at the print methods available to the designer of printed textiles and the different types of print designs in terms of motif, style and layout.

11.2 Initial considerations

Many print designers work for themselves as freelance designers. They may build up relationships with companies that they work for on a regular basis, producing designs to briefs set by the commissioning company, or they may build up a portfolio of work that they themselves take directly to companies or they sell through an agent. While an agent may take as much as 40% of the selling price of a design, this can still actually be the most economically viable way for freelance designers to sell their work. It allows them to concentrate on designing rather than selling, which can be a very time-consuming and costly activity.

Whether working to a brief as an in-house designer or freelance, or designing a collection for a personal portfolio that will be sold directly to companies or through an agent, print designers needs to be clear as to exactly what market area they are working in. They need to research predicted trends, themes and colours.

For print designers, it is important to know what market they are aiming at as the number of colours that they can use will depend very much on this. Fabrics that are going to retail at the top end of the market can have many more colours than fabrics at the cheaper end. They have to know how their design ideas would be interpreted into fabrics, and this means that they have to know what different printing methods are available.

11.3   Different classes of printing

While many different methods of printing have been developed, these can be divided into four different classes. They are dyed, resist, discharge and direct.

11.3.1   Dyed

In this method of colouring fabric, dyeing is used to colour parts of a fabric. A mordant, usually a metallic oxide, is used to act as a fixing agent for the dyestuffs. The mordant is painted on cloth in the desired pattern areas and the cloth is immersed in a bath of dye solution. The dye only becomes fixed to the fabric in those areas where the mordant has been painted because the mordant reacts with the dyestuff making it insoluble to washing and fast to light. Different mordants will react with a dye to give different colours and this means that a cloth painted with several mordants can be passed through one dyebath and rinsed off to give a pattern of many colours.

11.3.2   Resist

In the resist technique, a substance that prevents the cloth from taking up the dye (the resist) is applied to areas of the cloth. The cloth is then dyed. The areas treated with the resist are not coloured, retaining the colour of the ground or undyed fabric.

Along with mordant dyeing, this form of printing is older than discharge or direct. Early Coptic resist fabrics were created by using blocks to stamp on the resist pattern, the fabric then being passed through a dyebath of woad or indigo.

11.3.3   Discharge

In the early 1800s it was discovered that it was possible to print a fabric that had already been piece-dyed, with chemicals that would remove or discharge the dye. Printing such chemicals in selected areas allowed fairly intricate and fine patterns to be created and the areas where the colour was removed could subsequently be printed with a further colour. Nowadays colour discharges are produced using dyestuffs that are unaffected by the chemical discharging agent. Combining these with the discharge paste means that the original colour can be removed and a new one put in its place in the same printing process.

11.3.4   Direct

This is the direct printing of a colourant to the desired pattern areas. Until chemically produced dyestuffs were discovered, the only type of direct printing used pigments. Pigments merely coat the surface of a fabric, unlike dyes which penetrate and stain the actual fibres.

11.4   Printing processes and print types

11.4.1   Batik

This is one of the oldest forms of resist techniques. Wax is applied to the parts of a design that are not to be coloured, with a special tool called a tjanting. After dyeing, the wax is removed. For further colours the process can be repeated. A feature of many batik designs is a crackled effect. This is deliberately produced by crushing the cloth when the wax has hardened, causing the wax to crack and the dye to seep through these tiny cracks.

11.4.2   Tie dye

Tie dye is practised in many areas of the world. Cloth is tied or knotted in certain areas and then dyed. The dye cannot penetrate these tied areas and so patterns are formed. Stones and shells are often incorporated into the tying, and modern examples will use pegs and paperclips. The tying process can be repeated for subsequent colours.

11.4.3   Hand-painted mordanted cottons

Indian hand-painted cottons that used mordants and dyeing to colour the fabric were very popular in early seventeenth-century Europe. These fabrics were known as calicoes or chints; chint comes from the Hindu word meaning coloured or variegated, and is the origin for chintz, which now is generally considered to refer to a multi-coloured floral design printed on a white or ecru ground.

These fabrics are extremely important for several reasons. The designs were very popular and have a timeless quality that has led to these forming the basis for much dress fabric and furnishing design since the late sixteenth century, when they were first imported to Europe. The demand for these fabrics brought about the start of the European textile and dye industries.

The original Indian mordanted cottons were produced by hand. The cloth to be painted was first smoothed and burnished with buffalo milk to give it a fine burnished surface. Next, the design was outlined on paper, and transferred to the fabric by piercing holes through the paper along the main lines and rubbing charcoal through these. The outlines of red flowers and other black outlines were then painted in, using an appropriate mordant. The parts that were to be blue or green were left blank and the rest of the cloth was waxed. It was then dipped in a dyebath of indigo dye. The wax was removed by scraping and then washing. All the flowers and birds and stems were then painted in with several different mordants; acetates of iron, alum, chrome, zinc and tin. When dyed in a madder dyebath a rich range of colours — black, lilac, pink, brown, crimson and purple — was created. Sometimes, a further dyeing would take place. The dye was washed off from the unmordanted areas and the cloth was left in the sun and air to make the dyes fast. Finally small yellow areas were painted in. It is likely that this was saffron, but it is not known for certain as this dye was not fast and has nearly all disappeared from the surviving examples.

11.4.4   Block printing

Wooden blocks may be used to apply colourant directly to a fabric or to apply a resist.

Block printing was in use some 2000 years ago in India and China. In Europe, block printing is thought to have originated in the Middle Ages and there is evidence of early block-printed linen fabrics produced in Germany between 900 and 1300. Printed with pigments and with details painted on by hand, these early designs were often of animals enclosed in circular shapes.

Slow and labour intensive, block printing has not been used much since the 1870s. In the UK, it did continue on a very small scale for the production of expensive silk and fine wools with *Liberty's*, a famous print works, having a block printing department up until 1967.

11.4.5   Copper-plate printing

In block printing, the cloth receives its pattern from the pigment on the surface of the uncut areas of the block. In copper-plate printing, a copper plate with the pattern inscribed into it has the print colour applied to it. Excess colour is then removed from the smooth surface of the plate leaving the inscribed lines full of the colour that will be printed. Pressure is then applied to the plate, transferring the colour to the cloth. This type of printing is called 'intaglio'.

The advantages of using copper plates over blocks were that extremely fine lines were possible, tonal effects could be achieved and much larger repeats could be used. Plate prints were usually single colour. A difficult and slow process, it was awkward to produce smooth repeats and this meant that many of the designs were what is termed island or handkerchief designs.

11.4.6   Roller printing

Essentially, this is a mechanised version of plate printing. Instead of flat plates, the copper sheets which hold the dyes are formed into cylinders. On the printing machine, the cloth is taken around a large roller and the copper cylinders, one for each colour to be printed, revolve against this, having picked up the pigment or dye from passing through colour boxes. The surplus dye is scraped from the surface of the roller by a blade and this leaves colour to print onto the cloth only in the engraved portions. The cloth is passed over heated rollers to dry the dye, which is then fixed by steaming.

The main advantages of roller printing are its high production capacity and the excellent quality of fine line and tonal work that can be produced. However, there are some disadvantages: it takes a long time to change designs in production, engraving costs are high (which means long production runs are necessary to be economical), repeat sizes are limited by roller circumference, and there can be a reduction in the strength of colours. This last disadvantage is caused by what is called 'the crush effect'. This is where subsequent rollers both force colour printed by the previous rollers a little further into the cloth and remove a little of it. This dual forcing in and removal can lead to a reduction in colour strength of up to 50 per cent.

Roller printing is most suitable for woven fabrics; knitted fabrics are not usually roller printed as the rollers stretch the fabric, distorting the design and making registration of subsequent colours extremely difficult. Roller printing can produce almost any type of print design in terms of motif and pattern. It is a more suitable method than screen printing for designs where fine line detail is required.

11.4.7   Screen printing

The basic principle of screen printing is stencilling. A stencil is a thin sheet of paper or metal with areas cut out. Colour is applied and this passes through the cut areas but not the uncut areas. Early Japanese stencil prints were very intricate, with fine silk or human hair being used to give the stencils additional strength and support. The colour was passed through the stencils with a large soft brush.

Nowadays, a fabric, constructed to allow any applied colour to pass through it, is stretched across a frame. Areas are blocked out by some method to create what is essentially a stencil. When colour is applied the blocked areas do not allow the applied colour through and so create a pattern. Originally, the support fabric was silk and it was stretched on a wooden frame. This is why screen printing is often referred to as silk screen printing. Silk, however, absorbs moisture and this causes the screen to sag when used with water-based print pastes. The introduction of nylon and polyester has enabled the development of screen fabrics that maintain their tension when wet. Wooden frames are prone to warping when wet, so these have also been replaced, by metal frames.

Currently, screens are usually coated with a light-sensitive chemical. If certain areas of the screen are blocked, when the screen is exposed to ultraviolet light the chemical coating in these blocked areas does not react and can be washed out. In the areas that are exposed to the ultraviolet light, a chemical reaction takes place which causes the screen mesh to be blocked in these areas. Colour can be pushed through the unexposed areas of screen by means of a squeegee, which is a rubber-tipped piece of wood or metal, slightly narrower than the length of the screen.

Each colour requires a separate screen. Colours may be separated by hand or by computer to produce films with opaque areas where the colour is to print, and clear or translucent areas where the colour is not to print.

Up until the 1950s, most screen printing was done by hand. This was a slow and labour-intensive process. The development of automatic flatbed screen printing and rotary screens (mesh cylinders through which the dyestuff is squeezed) speeded up the printing process and it is rotary screen printing that is the most common type of fabric printing found today. Origination costs are relatively low compared to engraving, setting up time is shorter, the production rate is high, larger repeat sizes are available (although these are still limited by the circumference of the screen), and strong bright colours can be achieved.

Flatbed screen printing is used for tee-shirts, scarves, high-quality fabrics, and those with a very large number of colours and designs with very large repeats. Flatbed screen printing can be used for all fabric types including wovens, non-wovens and knits.

Rotary screen printing is used for a wide variety of products. The design size can be bigger than for roller printing, with a maximum design repeat of 90 centimetres, and a greater number of colours can be used — up to 16 compared to a maximum of 8 for roller prints.

11.4.8   Application prints

These are prints where the design is printed on a white or ecru fabric. The ground colour of the fabric will form an inherent part of the design.

11.4.9   Overprints

These are prints that are applied to previously dyed fabric. They are direct prints with the applied colour usually being darker than the background. Any print process may be used to apply the colour.

11.4.10 Blotch prints
In blotch prints both the background and motif colours are printed onto the fabric using a direct printing process. There can be a problem with designs that have large areas of ground in that the background colour can be uneven. Blotch prints that are printed with pigments can be rather stiff and boardy.

11.4.11 Devoré or burn-out prints
A base fabric that is made of two component fibres, one of which dissolves in a weak acid solution and one which does not, can be printed with a weak acid solution in selected areas. The fibre that dissolves will be removed from these areas, creating an engraved effect.

11.4.12 Discharge prints
Discharge prints may be white or coloured. White discharge prints are ones where the design motif is white, created by removing colour in that area from an already-dyed piece of fabric. Colour discharges are where many colours are applied to the discharged area.

Discharge prints are not widely found, as production is costly because of the necessity to dye and print. It is also a long and delicate process, and not getting it quite right can be expensive.

11.4.13 Flock prints
These are prints where adhesive is directly printed onto the fabric in desired areas and then short fibres are applied, which adhere only to these areas. It is usual for both base fabric and fibres to be dyed before flocking.

11.4.14 Transfer printing
Transfer or sublistatic printing was developed in the late 60s/early 70s. The pattern is first of all printed onto paper with special inks containing dispersed dyestuffs. These inks sublime at temperatures between 160 and 200 °F. Sublimation is a physical process whereby a solid is converted to a vapour by heat and back again to a solid on cooling. When heated, the inks have little affinity for the paper carrier but a high affinity for the fabric to be printed and therefore transfer to the fabric. The dyestuff is fixed by this process so no further finishing is required.

Transfer printing is a relatively easy and economical printing process. It is particularly suitable for unstable fabrics such as knits, where inherent stretchiness causes problems with accurate colour registration. Transfer printing allows a sharply defined print on knitted fabrics as all the colours are printed at once. The process is not a skilled one and, as no dye or finishing is required, it is a very clean process.

11.5   Developing design ideas

With the briefing meeting establishing what it is that is required, the market aimed at and the production processes that are to be used, the designer can start to develop design ideas.

Print designers usually work on paper, maybe developing ideas from their sketchbook. They will work out design ideas, giving consideration as to how the design might repeat (see Chapter 4 and below) and any restrictions with regard to repeat size that might be imposed by the product being designed for. Some designers will work at a very early stage in the design process on a CAD system, using cut and paste and other such facilities to more easily manipulate their chosen imagery.

However a print designer works, designs for printed textiles are first of all fully worked-out on paper. This artwork is subsequently reproduced by superimposing the pattern onto an already existing fabric using one of the printing processes previously described. Any medium can be used in the artwork, but materials such as designer's gouache or inks, with which designers can achieve good representations of particular effects in manufactured fabrics, allow design problems to be more accurately resolved in the studio before the design is put into production. CAD systems are being used more and more in the preparation and development of designs for printing. These systems allow designers to scan-in images and easily manipulate them.

11.6   Classifying printed textile designs

Print designers may be asked to produce designs with particular motifs or within a specified subject area or theme, in a particular style and in a particular arrangement or layout, and it is helpful therefore to be familiar with how printed textile designs are described.

Designs may be described by

1) the motif or subject matter,
2) the style in which they are rendered, or
3) the arrangement or layout of the motifs.

These three elements, combined with the way a design is coloured, are the essentials that form a print design.

11.6.1   Motifs and styles

11.6.1.1   Florals
The most common and best-selling prints in both apparel and furnishing fabrics for the past several hundred years have been florals. These are not monotonous motifs and they have no negative connotations. Flowers may be represented realistically or stylised. Whatever design style is popular at a particular time, some interpretation of plant form is likely to be seen.

Chintz, which was the term used for Indian mordanted cottons, is now the name usually given to medium-scale, all-over floral patterns. The colours are usually in fairly strong and vibrant mid-tones on a white or ecru ground.

The floral designs of the Arts and Crafts movement typified by the work of William

Morris are usually winding designs where symmetry plays an important part. Many of these were taken almost directly from Persian designs, with little change or alteration, and are often described as Arts and Crafts florals.

Realistic representations of flowers and plants in designs are often classed as botanical. Often the motifs are frequently repeated as entirely separate elements.

Small-scale, simple floral designs that are regularly repeated are often referred to as Provençal, after Provence in France, where such designs were block printed in the eighteenth century.

11.6.1.2   Pictorial and figurative designs
Other natural forms such as shells, animals and landscapes are traditionally popular motifs in textiles.

Many historical landscapes or scenic patterns were printed at a famous print works at Jouy in France. Such designs have become known as 'toiles de Jouy'. They are usually printed in one colour on a natural (undyed) ground; the original fabrics used the Copperplate process.

Other pictorial or figurative designs are inspired by artefacts — coins, plates, etc. Often such pictorial designs are amusing and use strong images. In the 1950s there was quite a fashion for pictorial designs, with kitchen fabrics incorporating vegetables and kitchen utensils as motifs.

11.6.1.3   Paisleys
These designs are based on original motifs from India and are often derived from the tree of life. The name comes from the Scottish town of Paisley where these motifs were woven into scarves and shawls, and the curved tear-drop shape that occurs so frequently in these designs is now known as a paisley.

11.6.1.4   Geometrics
These designs use geometric shapes. Much Islamic pattern is geometric as, for religious reasons, people and animals could not be portrayed.

11.6.1.5   Abstracts
Abstracts are considered to be any non-figurative design. The most important features are the shapes and colours.

11.6.1.6   Ethnic designs
Ethnic designs are those designs inspired by traditional design work from different native cultures — African, South American, North American Indian, etc.

11.6.1.7   Co-ordinating designs
Co-ordinates are designs developed to be used together. Sometimes, these will use similar motifs but on different scales; usually the same colour palette will be used throughout. Groups of co-ordinates are often developed and presented as a 'story'.

11.6.2   Layouts
The layout of a design is the arrangement of motifs in the framework of the design plane.

11.6.2.1   Tossed patterns

A tossed pattern is one in which the motifs in a repeat do not occur at regular intervals. It is as if the designs are dropped in a fairly random way onto the fabric. Such designs may be spaced with ground area between motifs, or packed so that the motifs touch, showing hardly any ground.

11.6.2.2   All-over designs

An all-over design has balanced motifs that recur regularly within the repeat unit. The motifs cover the fabric with little ground showing.

11.6.2.3   Foulards

Foulards are patterns with small motifs repeated directly under and across from one another at measured intervals. The name comes from the French word for a scarf or necktie.

11.6.2.4   Ogees

The symmetrical, onion-shaped layouts often used in William Morris's designs are called ogee layouts.

11.6.2.5   Stripes

Printed horizontal stripe patterns are called bayadéres. Vertical stripes are not really possible in flatbed screen printing because the joins will tend to show. Printed diagonal stripes are usually at a 45° angle. This is for ease in matching up the design when pieces have to be joined.

11.6.2.6   Borders

Border patterns are those where the design is focused along one selvedge. Some printed designs have borders along both selvedges.

11.6.2.7   Engineered designs

An engineered design is when one screen makes the entire pattern. Examples of engineered designs include towels, rugs, duvets and headscarves.

11.6.3   Pattern direction

When developing designs for printing, the direction of motifs must be considered. One-way patterns are where all the motifs point in the one direction. In apparel, this can cause problems when laying-up ready for cutting, as all the pattern pieces have to lie in the same direction and this will normally result in fabric wastage. Two-way patterns are those where some motifs go one way while others lie the opposite way.

One-way and two-way prints are both directional. Non-directional prints are where the motifs face in all directions; there is no top or bottom to the design. One-way patterns are often used for upholstery or curtaining fabrics. In two-way or multi-directional patterns, the images must work in different directions.

11.7   Design size

The size of a design is determined by several factors — the size of motifs and layouts, the desired end-use of the fabric and the production equipment being designed for.

Apparel fabrics are usually designed in croquis form, with repeat sizes varying enormously. Upholstery and curtain fabrics, however, have standard repeat sizes so they are usually designed in repeat. It is important to consider the scale of motifs as they relate to end product.

11.8   Repeats and colourways

While for many print designers their involvement may end with the production of a design on paper, others will have to ensure that their design is drawn in repeat and develop colourways. Some of the commonest types of repeats used in print design are block or simple repeats, half drops, tile or brick repeats, and ogees. Mirror imaging is also important in print repeating patterns. (See Chapter 4 for a fuller description of repeats and repeating.) Colourways will often be developed using a CAD system. It is a simple process to scan in a design and then use graphic software to try out different colourways. Both balanced and unbalanced colourways might be developed. The overall appearance of a design can change quite dramatically by altering colour balance within the design.

11.9   Base fabrics

The base fabric on which a design is printed will have an impact on how the final design appears — whether it is lightweight or heavyweight, whether it is coarse or fine, thick or thin, opaque or transparent, a pile fabric or smooth — and print designers should have a knowledge of the different types of woven and knitted fabrics that they can print on. More important, however, is the fibre composition of the base fabric. This will affect the different types of dyestuffs that can be used to colour the cloth. Table 11.1 shows the main fibre types, and the dyestuffs that can be used to dye these with their main characteristics.

11.10   Dyes and pigments

However a fabric is to be coloured and whatever dyes are to be used, it should be scoured to remove any impurities that have been introduced during weaving or knitting.

The choice of colourant will largely depend on the fibre content of the base fabric and the fastness requirements. Dyeing is a chemical process which involves the migration of the dye from the dye solution to the surface of the fibres in the fabric being coloured, the diffusion of the dye through the fibres and the fixing of the dye by chemical bonding. Printing with pigments involves a chemical binder being first mixed with the pigment, and it is this binder which combines with the fabric rather than the pigment itself.

11.11   Print sampling

Many sample fabrics are made by hand-screen printing, so that design and colourways

Table 11.1   Dyes, their uses and their fastness.

Type of dye	Fibres able to be dyed	Fastness to light	Fastness to washing	Fastness to crocking (transfer of colour when rubbed)	Fastness to bleach
Acid	nylon wool	good	varies	good	fair
Azoic	cellulosic	excellent	excellent	excellent	excellent
Basic	acrylic some nylon some polyester	varies	excellent	excellent	poor
Disperse	acetate acrylic nylon polyester	good	good	good	good
Mordant	cotton wool	varies	fair	fair	varies
Pigment	all fibres	excellent	fair	fair	excellent
Reactive	wool	good	excellent	excellent	poor
Vat	cellulosic	excellent	excellent	good	excellent

can be checked before full production commences. This is a relatively easy process, and the only real restrictions are in the number of colours in terms of budget and the repeat size. The design must be drawn accurately in repeat and the edges of the design must be disguised by using what are termed 'lines of least resistance'.

The colour separations may be made by using opaque paint to paint by hand a separate film for each colour. These films or positives are then checked on a light-box to make sure there are no gaps or pinholes. The positives are then attached to the screens, coated with the light-sensitive chemicals and then exposed to ultraviolet light. The areas of polymer solution not exposed to the light will wash off and these are the open areas of screen through which the print paste will pass. This process is carried out in a dark room. After washing the screen will be checked for any pinholes or faults which need to be repaired.

With the screens prepared, recipes for the required colours have to be established and the pigments or dyes to be used made up. A first sample, called a 'strike-off', will be printed. From this the colours can be checked, as can the registration of the screens — to see that the colours of each screen fall in just the right places. Colour recipes may have to be altered and the printing table adjusted to ensure best reproduction of the design.

When the strike-off is accepted by the designer, the sample fabric can be printed. After printing, the fabric will be dried, steamed and any excess pigment or dye washed off.

11.12   Making particulars

Again, it is often the responsibility of the designer to ensure that accurate records are kept to ensure that the production office will be able to exactly re-create the samples approved. They will keep information on the base cloth (its weight, composition and structure), the design, and the dyestuffs and recipes for the colours required. Any applied finishes will also be recorded.

11.13   Summary

Printing is rather an ambiguous term. Although many methods of colouring fabrics are described as printing, strictly speaking many of these are dyeing techniques. There are several elements that go together to make up a print design; the base fabric on which the pattern is made, the design and the way the unit repeat of this is repeated across the fabric, and the types of dyestuffs applied and the way these are applied.

The main methods of printing are batik, tie dye, hand painting with mordants, block printing, screen printing, roller printing and transfer printing, and these can be divided into four classes – dyed, resist, discharge and direct. Application, blotch and devoré are all different types of prints that use direct printing.

Print designs can also be classified by motifs and style, by the way these are laid out to repeat across the fabric and print type.

Most printed textiles are first designed on paper. Any medium can be used, but materials with which designers can achieve good representations of particular effects in manufactured fabrics allow design problems to be more accurately resolved in the studio before the design is put into production. As computer-aided design (CAD) systems continue to develop, these are being used more and more in the development and preparation of designs for printing.

Bibliography

Colchester, C., *The New Textiles: Trends and Traditions*. London, Thames & Hudson, 1993.

Harris, J., *5000 Years of Textiles*, British Museum Press, London, 1993.

Hatch, K.L., *Textile Science*, New York, West Publishing Company, 1993.

Meller, S., *Textile Designs: 200 years of Patterns for Printed Fabrics Arranged by Motif*, London, Thames & Hudson, 1991.

Paine, M., *Textile Classics: a Complete Guide to Furnishing Fabrics and their Uses*, London, Mitchell Beazley, 1990.

Storey, J., *The Thames and Hudson Manual of Dyes and Fabrics*. London, Thames and Hudson, 1978.

Storey, J., *The Thames and Hudson Manual of Textile Printing*, rev. ed., London, Thames and Hudson, 1992.

Yates, M., *Textiles: a Handbook for Designers*, rev. ed.. New York; London, W.W. Norton, 1996.

Appendix A

Sample Gantt chart for a textile design project

A stylist working for a towelling manufacturer has been asked to put together a range of embroidered towels for a store group. There should be three different designs and these are to co-ordinate with a range of 12 plains currently selling well.

Steps required to complete the project.
Project briefing then:

Initial research to include	visit to major trade fair
	shop visits to see what is currently on offer
	look at forecasts and predictions in this and related product areas
Design idea generation	brainstorming for suitable themes and sources
	initial paperwork ideas
Design development	development of ideas on paper and using CAD system
	initial embroidery sampling and colourway development
Range selection	selection and sample production of proposed range and colourways
Range presentation	presentation of proposed towel designs with colourways and full making particulars

Table A.1   Example Gantt chart.

	Week 1	Week 2	Week 3	Week 4	Week 5	Week 6	Week 7	Week 8	Week 9	Week 10	Week 11	Week 12
Initial research	▓	▓										Range presentation
Idea generation			▓	▓	▓							
Design development					▓	▓	▓	▓				
Range selection										▓	▓	

Appendix B

Some tips for presenting work

Measuring

Measurements should be accurate. Time taken to measure with a ruler will save time in re-sticking and re-mounting work. If a piece of artwork is intended to be in the middle of a board, then measure to ensure it is. If a right angle is intended, then make sure it is a right angle.

Cutting

Sharp scissors and cutting knives should be used. The modelling knives that have blades with a series of removable sections are excellent. Ragged cutting lines can be avoided by cutting on a suitable surface. Where available, use a guillotine.

Paperwork that is deliberately torn should look as if this is intended and not that the mounting is just careless.

For cutting fabrics, sharp scissors are a must. To prevent the edges fraying, masking tape can be used as a backing round the fabric edges. Woven fabrics are best cut along ends and picks. Fabric should be ironed or pressed first, as squarely as possible. Beware — pinking shears can give some strange corners to fabrics.

Mounting

Be creative in what work is mounted on, although whatever material is used, it must be appropriate to the type of work being mounted. A wide variety of materials can be used including corrugated cardboard, blotting paper, brown wrapping paper, sugar paper, and even old books and magazines. Heavy cartridge paper in A4, A3 and A2 pads is a simple and effective base for much design work.

Ecru, black and grey backgrounds often work well as they interfere less with the colour of the design work itself. Bright mounts can overpower the work being presented although there are of course times when they are appropriate. Shiny mounting boards can be difficult to use, picking up fingerprints and other marks easily.

Gluing

When sticking anything down, it must be done neatly and all the edges of any paperwork must be firmly fixed. Be careful that glue does not stretch paperwork: particularly watch cartridge paper that has not been stretched. See Table B.1 for a summary of hazards to watch out for.

Table B.1   Hazards to watch out for when gluing.

Type of glue	What to watch for
Spray mount	Drops off after a while, health hazard (should use a mask when spraying), can be messy — drifts to areas other than that being sprayed
PVA glue	Can be messy
Cow gum	Very messy although it can be rubbed off some surfaces
Glue stick	Can stretch thinner papers, difficult to move once stuck down
Double-sided sticky tape	Very good for most sticking jobs, difficult to move once stuck.

The use of masking tape on the back of paperwork before applying a glue stick or double-sided sticky tape enables you to take off the artwork without wrecking it.

Finishing

All charcoal and pastel drawings should be sprayed with fixative (hair spray is a good, if smelly, fixative) and trimmed. Where possible, any group of such drawings should be trimmed to a common size.

The best design work is often simple. This is also true for presenting work — keep it simple — allow the design work to be seen. Negative space is important; space around artwork can be very effective.

Appendix C

Example of a simple structure for letters

1, The Address
With town
County
And POST CODE

Mr J Smith
2, Address
Againwithtown
County
And POST CODE

Date

Dear Mr Smith,

Re: A Simple Layout for Letters

As promised I am writing with a suggested simple layout for letters. This outline is a good base and starts with a polite introduction.

If the person you are writing to is not known to you but you know their name, then you should start 'Dear Mr' (or other such title). If you do not know the name but are writing to a position, for example the Managing Director of a company, then you should start 'Dear Sir' or 'Madam' as appropriate. Likewise, at the end, if you know the name of the person you are writing to you can finish with 'yours sincerely'; if you do not know the name then 'yours faithfully' is more usual.

Your letter should be structured in a logical manner and should be easy to read, pleasant and clear. It should be as short as possible while covering the subject matter adequately. Remember to include the date with the full address for both the sender and the addressee. It is also always worthwhile to spell-check all your work to prevent silly errors creeping in, and you should make sure you read the whole letter through when finished to ensure that it reads well and will make sense to the reader.

For a designer particularly, layout is important. A potential client may lose confidence in a designer's capabilities if that designer cannot solve the visual design problems posed by a letter.

As with the introduction, the final paragraph should conclude in a polite and friendly manner as I am about to do.

I hope this has been of some help but please do not hesitate to contact me again if I can be of any further assistance.

Yours sincerely,

Jacquie Wilson (Ms)

Appendix D

Example fabric specification sheet for a woven fabric

Weave specification sheet

Loom	Number of shafts	Reed	Width on loom
Ends per 10 cm (loom)	Picks per 10 cm (loom)	Weight per square metre (loom)	Weight per square metre (finished)
Ends per 10 cm (finished)	Picks per 10 cm (finished)	Description/reference number	

Count & quality	Colour	Warp pattern

Count & quality	Colour	Weft pattern

Draft

Design

Lifting plan

Appendix E

Example fabric specification sheet for a knitted fabric

Knitted fabric specification

Machine type	Gauge	Weight per square metre (finished)
Wales per 10 cm (finished)	Courses per 10 cm (finished)	Description/reference number

Colour	Count & quality	Number of courses																	

Loop diagram

12

11

10

9

8

7

6

5

4

3

2

1

24

23

22

21

20

19

18

17

16

15

14

13

Appendix F

Calculating percentage compositions

When several different yarns with differing fibre compositions are combined in one fabric, the resulting fibre composition for the new fabric has to be calculated. This will often be done by the production office, although sometimes it may be the job of the design studio to work it out. In either case it is important for designers to understand the procedure.

Example:
A fabric is made from three yarns.

> Yarn (a) is a cotton/acrylic blend — 50% cotton/50% acrylic
> Yarn (b) is a linen/cotton blend — 60% linen/40% cotton
> Yarn (c) is 100% acrylic

A piece of the fabric weighs 400 grams.
    To work out the percentage composition of the fibres in the fabric it is necessary to unrove the fabric. The resulting unroved yarns are weighed as yarns (a), (b) and (c). It is found that

> Yarn (a) weighs 175 grams
> Yarn (b) weighs 125 grams
> Yarn (c) weighs 100 grams

From this information we can work out the percentage composition of yarns in the fabric and therefore the percentage composition of the fibres.
    The percentage composition of yarns in the fabric:

$$\text{Yarn (a) as a percentage is } 175 \times \frac{100}{400} = 43.75\%$$

$$\text{Yarn (b) as a percentage is } 125 \times \frac{100}{400} = 31.25\%$$

$$\text{Yarn (c) as a percentage is } 100 \times \frac{100}{400} = 25\%$$

The percentage composition of fibres in the fabric:
43.75% of the fabric is made up of yarn (a). This yarn is 50% cotton/50% acrylic

$$\% \text{ cotton} = 50\% \text{ of } 43.75\% = 21.87\%$$

$$\% \text{ acrylic} = 50\% \text{ of } 43.75\% = 21.87\%$$

31.25% of the fabric is made up of yarn (b). This yarn is 60% linen/40% cotton

$$\% \text{ linen} = 60\% \text{ of } 31.25\% = \frac{60}{100} \times 31.25 = 18.75\%$$

$$\% \text{ cotton} = 40\% \text{ of } 31.25\% = \frac{40}{100} \times 31.25 = 12.50\%$$

25% of the fabric is made up of yarn (c). This yarn is 100% acrylic

$$\% \text{ acrylic} = 100\% \text{ of } 25\% = 25\%$$

To get the percentage composition of the fibres in the fabric we must add together the percentage compositions as established:

$$\% \text{ cotton} = 21.87\% + 12.5\% = 34.37\% = 34\%$$

$$\% \text{ acrylic} = 21.87\% + 25\% = 46.87\% = 47\%$$

$$\% \text{ linen} = 18.75\% = 19\%$$

to the nearest whole number

Alternatively, the fibre composition can be worked out as below:

Yarn (a) weighs 175 grams and is 50% cotton/50% acrylic;
therefore weight of cotton is 50% of 175 grams = 87.5 grams
and weight of acrylic is 50% of 175 grams = 87.5 grams

Yarn (b) weighs 125 grams and is 60% linen/40% cotton;

$$\text{therefore weight of linen is } 60\% \text{ of } 125 \text{ grams} = \frac{60}{100} \times 125 = 75 \text{ grams}$$

$$\text{and weight of cotton is } 40\% \text{ of } 125 \text{ grams} = \frac{40}{100} \times 125 = 50 \text{ grams}$$

Yarn (c) weighs 100 grams and is 100% acrylic;
therefore weight of acrylic = 100 grams

The weights of the fibres are therefore:

cotton — 87.5 grams + 50 grams = 137.5 grams

acrylic — 87.5 grams + 100 grams = 187.5 grams

linen — 75 grams

and the percentage compositions are:

$$\% \text{ cotton} = 137.5 \times \frac{100}{400} = 34.38 = 34\%$$

$$\% \text{ acrylic} = 187.5 \times \frac{100}{400} = 46.88 = 47\%$$

$$\% \text{ linen} = 75 \times \frac{100}{400} = 18.75 = 19\%$$

Appendix G

Getting press coverage

There is a certain technique involved in getting journalists to take stories. First of all, the appropriate magazines and papers (those that are likely to want to use the story) should be researched. Something interesting enough to attract the attention of the editor is necessary: an unusual slant if possible about a successful job, the experience of setting up in business, etc.

The normal way to approach the press is to write a press release and send it to selected magazines and papers, with professional photographs. Releases should be followed up by phoning the editors/journalists. Samples may be requested for photography for a feature. The deadline that may have to be met may seem a nuisance; it will, however, most certainly be worth it if an article is run. The story may not be taken up immediately but kept on file until it can be used in a suitable edition or feature. Most magazines work to deadlines 1–3 months in advance of publication.

Appendix H

A structure for fee letters

Structure

Subject heading
Introduction
Summary of the brief
Services provided

       Stage 1
       Stage 2
       Stage 3
       Fees for stage 1—
       Fees for stage 2—
       Fees for stage 3—

Special clauses
Additional fees
Fee instalments
Expenses chargeable
Request for written acceptance
Conclusion

Subject heading

A heading at the beginning of the letter makes the reader clearly aware of what the letter is about.

Introduction

The introduction should be polite. It is courteous to say thank you for an interesting briefing meeting, lunch and visit, and to express interest in the job. It is practical to refer to the date and place of the briefing meeting and to any other persons who may have been present.

Summary of the brief

It is better to err on the side of too much detail rather than too little. This section allows the designer to set down his understanding of what is required. It is better that any misunderstanding of a client's requirements on the part of the designer is cleared up before any time is spent on design work that would not be answering these requirements. If the client has been vague about any areas such as production costs or budget, then this can be tactfully recorded here.

Services provided

Stage 1

Discussion with specialists; number of visits; preparation and submission of preliminary work. Discussions with specialists and technicians might refer to visits to showrooms, factories, branch offices, etc.

With regard to preliminary designs, it is useful to try to state the number that will be given. Most clients would expect to see several alternative preliminary designs so that they could have some say in the ultimate choice and thus feel they were getting value for money. The form that this preliminary design work will take should also be given, i.e. sketches, fabric samples, mock-up garments, etc.

Stage 2

Development of work. The client will either approve the design ideas submitted or request some alterations. As a result, the designs will either be developed further or amended. If there is no provision in the fees for amendments then a special clause would normally be included to allow for any time that may be required for this.

Stage 3

This depends on the type of design work. Finished artwork may involve putting designs into repeat, transferring the design to point paper, working out alternative colourways, inspecting strike-offs and first production runs, etc.

Special clauses

Copyright

A designer can retain full copyright, convey restricted copyright to the client or convey full copyright. A printed textile design for curtains may be sold with restricted copyright to prevent it being used as, say, a wallpaper design without further fees to the designer. Few clients will, however, agree to anything other than them having full copyright (but this should not be granted until fees have been paid in full).

Breaking clauses and fees for abandoned work

If a client does not decide to continue with a project, for example because company policies change or the company is taken over, then it is not unreasonable to expect to be paid for work done. This should be made clear in the fee letter.

Permitted modifications

The Chartered Society of Designers advises that clients should not make alterations to designs without the consent of the designer. This applies when copyright is retained until the payment of final fees.

Samples for records

A tactful claim for samples of work done, as appropriate, should be made in the fee letter for the designer's records and portfolio.

Signed work and design credits

Designers are entitled to claim authorship of a design. A designer's consent must be obtained before his name or signature is reproduced. This is a matter for negotiation and would not normally be referred to in a fee letter.

Additional fees

Extra visits

It is worthwhile putting something in the fee letter to say that any visits extra to those allowed for in the fees quoted would be charged separately.

Changes to brief

Again something should be included to the effect that any changes to the brief that mean more work will be charged accordingly, probably at an hourly rate.

Fee instalments

Rather than wait until the end of a long job before sending an invoice, it is much better to invoice at the completion of each stage. This should be indicated in the fee letter. Any fees charged at an hourly rate in addition to the fees for the stages should be invoiced monthly and the client should be told of this.

Expenses chargeable

If not taken into account in the fees for the different stages, expenses for travel, etc., would be invoiced in addition, and the client must be told this.

Request for written acceptance

Confirmation, in writing, of the client's acceptance of the fee letter must be asked for. It is usual to ask that they sign and return a copy of the fee letter. No design work should ever be started until written acceptance is given by the client.

Conclusion

A polite conclusion asking that the client will let the designer know of any queries is normal.

Appendix I

Sample fee letter

**The Design Consultancy
65 New Road
Old Town
Lancashire
LA1 2AA**

Mr J Smith
The Retailer in the High Street
26 High Street
London
L22 6TT

November 30, 2000

Dear Mr Smith,

Re: Design of Range of men's casual/sporty knitwear for Autumn/Winter 2001/2002

Thank you for your hospitality when I visited last week. The project we discussed is a very interesting one and, as agreed, I am now writing to set out what I could do for you and what my fees would be.

The brief is to design six styles for a range of men's 7–5 gauge knitwear for Autumn/Winter 2001/2002. Each style should have two alternative colourways. The range is to be aimed at a relatively sophisticated young man and would retail at around £45. Targeted primarily at the UK market, the styles may be sold in your European stores, although this is not critical.

Stage 1

This would involve some initial market research to look at what is available currently and to identify trends for 2001/2002. A further visit to your main office would also be necessary to find out more about where you outsource your production.

Stage 2

This comprises initial design work to include mood boards and lifestyle boards and some initial fabric sampling with styling ideas. This stage should be completed and presented to you by the end of February 2001.

Stage 3

Following a mid-project review meeting, initial design ideas would be developed into sample garments in conjunction with agreed manufacturers. The range of sample garments should be ready for presentation by the end of March.

My fees for stages 1 to 3 above are as follows:

Stage 1:   £672
Stage 2:   £1,220
Stage 3:   £1,220

Fees for each stage will become due on completion of that stage. If, for any reason, the project is terminated by yourselves, then fees would be due up to and including the stage of work being undertaken at the time of termination.

Travel expenses for visits to your head office and any other required travel will be invoiced separately. Any such travel will be agreed with yourselves in advance.

Any additional work to that set out above, including any checking of sample production as necessary, will be charged at my hourly rate of £32.00.

It would be appreciated if I could have, for my records, samples of all fabrics that go into production.

There are two copies of this letter enclosed and I would ask that you sign one and return it to me to indicate acceptance of the terms of reference as set out in this letter.

If there are any queries with regard to my understanding of the project and how I intend to progress this, please do not hesitate to get back to me. I am very much looking forward to starting work on this project and to hearing from you.

Yours sincerely,

Ania K Designera

Appendix J

Calculating an hourly rate

Assume a salary of £15 000 per annum. With an 8-hour day and a 5-day working week, there are 2080 hours in a year.

$$\frac{£15\ 000}{2080} = £7.21\ \text{per hour}$$

There is, however, no way that 2080 hours can be sold to clients. Allowing 8 days for public holidays, three weeks for personal holidays and one week for unforeseen illness is not unreasonable.

8 days public holiday (8 × 8)	= 64 hours
3 weeks personal holiday (15 × 8)	= 120 hours
1 week for illness (5 × 8)	= 40 hours
Total	= 224 hours
2080 − 224 = 1856 hours	

Now there is cover for any time off that is taken.

$$\frac{£15\ 000}{1856} = £8.08\ \text{per hour}$$

Overheads, however, have not been taken into account. Overheads represent all the money that has to be spent in order to be able to sell time. They would normally include rent, rates, insurance policies, heating, lighting, telephone, postage, typing, stationery, drawing materials, accountant's fees, bank charges, advertising, and membership of professional bodies. While a few will be known, some will have to be guessed/estimated. Overheads would usually be worked out as an annual figure.

Any one-off starting-up expenses would have to be taken account of as well. These are the one-off expenses incurred in starting up a business. Initial legal expenses, office furniture and any equipment needed would come under this, and would be termed capital outlay. These expenses would be added together and, if it were felt to be reasonable that they should be recouped in the first year's fees, they would be added to the overheads. If it was felt more reasonable to account for them over two or three years, then a smaller percentage would be added.

An estimated figure for overheads and capital outlay might be £10 000. These costs cannot be sold directly to clients. They must be sold indirectly by spreading the total over every hour that is to be sold.

$$\text{Cost of one hour} = \frac{\text{salary + overheads + \% for capital outlay}}{\text{hours worked}}$$

$$= \frac{£15\ 000 + £10\ 000}{1856} = £13.47$$

But this is still not the final selling cost.

A working year of 1856 hours has allowed for holidays but it is being assumed that every one of these 1856 hours can be sold to clients. This can never be the case, as time must be spent on work that cannot be directly charged to any one account, e.g. time spent filing, delivering, and visiting exhibitions and trade shows. A large proportion of unsaleable time must be given to administrative responsibilities, looking for clients, following up enquiries, running office finances and the like. This time will probably run to at least $\frac{1}{3}$ of working hours. Therefore available time remaining

$$= \frac{1856 \times 2}{3} = 1237.33 \text{ hours}$$

$$\frac{£25\ 000}{1237} = £20.21 \text{ per hour}$$

If all the 1237 hours could be sold at this rate, then the business would break even. That is, enough would have been made to pay the basic salary and cover overheads and capital outlay. There has, however, been no profit made. The profit is the reward for the worry and uncertainty of being in business. It is usual to add profit as a percentage, and 15% is a reasonable figure:

$$\frac{£20.21 \times 115}{100} = £23.24 \text{ per hour}$$

An appropriate hourly rate to base fees on is therefore £23.00.

Should circumstances alter from those predicted, then the hourly rate should be adjusted accordingly.

Summary

$$\text{cost of one hour} = \frac{\text{salary + overheads}}{\text{hours that can be sold}} + \text{\% for capital outlay + \% profit}$$

Glossary

Acid dyes Dyes used for dyeing nylon and wool.

Additive colour A mixture of coloured lights. The three primary colours of red, green and blue, when mixed together in equal proportions, produce white light. Mixing the three additive primaries in differing amounts can create any colour in the rainbow. Colour televisions use the principle of additive colour mixing.

All-over designs Designs with balanced motifs that recur regularly within the repeat unit. The motifs cover the fabric with little ground showing.

Apparel textiles The clothing or apparel market includes most garments that are worn.

Application prints Prints where the design is printed on a white or ecru fabric. The ground colour of the fabric forms an inherent part of the design.

Azoic dyes Dyes that are produced directly in the fibre by the chemical combination of their constituent parts. These are used for cellulosic fibres and have excellent fastness.

Balanced colourways When the colours change but tonal relationships of the colours within the design stay the same, giving the same overall visual effect.

Bayadéres Printed horizontal stripe patterns.

Blotch prints Prints where both the background and motif colours are printed onto the fabric using a direct printing process.

Body blanks Knitted panels with integral ribs, which are then cut and sewn into garments.

Bonded fabrics Fabrics produced by bonding webs of fibres together by stitching or by adhesive.

Borders Designs where the pattern is focused along one selvedge or other edge. Some printed designs have borders along both selvedges.

Brainstorming In its most formal sense, this is a group participation technique for generating a wide range of ideas to tackle a stated problem. In a less formalised way, an individual or pair can use a brainstorming session to generate ideas.

Brand A trade name identifying a manufacturer or product.

Brief Describes a design project.

Briefing meeting Where the work that is required is established/when the designer finds out what is required of them.

British Knitting and Clothing Export Council Organisation that aims to improve the export performance of the apparel/fashion industry by assisting both existing and potential exporting companies with marketing and promotional activities and by creating a greater overseas awareness of the British fashion industry.

Burn-out prints Prints with an engraved effect made by dissolving out one component fibre from a base fabric made with two distinct fibre types.

Business plan A plan for a proposed business which includes projected sales.

Buyers and selectors The people responsible for deciding which styles are going to be available as part of a product range.

Carriage Yarn-carrying device on a knitting machine.

Centring When a fabric design is organised in such a way that it is balanced about the middle line of a fabric in a vertical direction.

Chartered Society of Designers (CSD) A UK professional body representing designers from all areas of design. Its aims are 'to promote high standards in design, foster professionalism and emphasise designers' responsibility to society, clients and other designers'.

Chroma Refers to the saturation of a colour or its colour strength.

Circular knitting Fabrics knitted on machines that have the needles arranged in slots around the circumference of a cylinder. The fabric is manufactured as a tube.

Colour and weave effects Patterns that are created in woven fabrics by combining colour with weaves.

Colour forecasting The selection of ranges of colours that are deemed to be those that will be wanted for a particular product/market at a particular time in the future.

Colour palette A range of selected colours that will usually consist of groups of colours, chosen with regard to trends and predicted directions.

Colour recipe List of component chemicals and pigments or dyestuffs with relative quantities required to produce desired colour.

Colour separation In screen printing, a separate screen is required for each colour within a design. Colour separation is the process whereby individual films are made for each colour. Colours may be separated by hand or by computer to produce films with opaque areas where the colour is to print, and clear or translucent areas where the colour is not to print.

Colourists Designers who predict colour trends and put together palettes of colours for specific seasons and product groups. Other colourists will work further down the design process line, colouring work produced by other designers to create different and alternative colourways.

Colourway Alternative colouring of a design, on paper or in/on fabric.

Computer-aided design (CAD) systems Computer software that has been developed to assist in the design process.

Concept selling This is where products are sold on a lifestyle basis.

Constructed textiles Includes woven textiles, knitted textiles, lace and carpets.

Constructive avoidance Time spent on work that is neither important nor urgent, in preference to urgent and important work.

Consultant designer Designer employed by a company usually on a part-time basis, to advise on design matters.

Consumer textiles Textiles not falling into the categories of apparel, furnishing, household and industrial; tents and back packs may be referred to as consumer textiles, awnings, umbrellas and luggage are also often classed in this category.

Contract furnishings Furnishings used in offices and public buildings such as schools, hotels and hospitals.

Converters Organisations that buy grey cloth and convert this by having it dyed or printed and then finished.

Co-ordinating designs Designs developed to be used together. Sometimes these will contain similar motifs but on different scales; usually the same colour palette will be utilised throughout. Groups of co-ordinates are often developed and presented as a 'story'.

Copper-plate printing A method of intaglio printing that uses engraved copper plates.

Copyright Exclusive right, granted by law, to control the use of an original piece of artwork or design.

Corporate identity Company image.

Cost price That price that essentially will cover any costs involved and ensure that no loss is made.

Cottage industry Domestic system of manufacture where textiles were spun, knitted and woven in the home.

Courses Horizontal rows of stitches in knitted fabric.

Cramming Where more threads are packed into some areas of a fabric than others.

Crochet A way of making fabric by looping and intertwining thread with a hooked needle.

Crocking The transfer of colour by rubbing.

Croquis A balanced design for print (*croquis* is a French word meaning sketch). A croquis should give the impression that would be seen if a frame were placed over any section of the finished cloth. Although not in repeat, a croquis will give the feeling of being in repeat.

Crossings Sections where warp and weft patterns combine more by chance than by design.

Cup-seaming A type of sewing that joins edges (usually two selvedges) by means of a chain stitch.

Cut and sewn knitwear Garments made from pieces cut from panels, or lengths of fabric, usually with integral ribs.

Cylinder The needles that are arranged vertically round a circular knitting machine; equivalent to the flat machine's front bed.

Decorex Furnishing show in London.

Denting In weaving, the way warp threads are arranged in the reed.

Dents In weaving, the spaces in the reed through which the yarns are threaded in groups of two, three or more, depending on the number of ends per centimetre required, the type of yarn being used and the weave.

Design Council The UK's national authority on design. Its role is to advise and influence business and others on the importance of design in terms of its contribution to the UK's international competitiveness.

Design Intelligence Organisation publishing trend and colour forecasts.

Design Management Institute (DMI) Organisation which aims to 'inspire the best management of the design process in organisations world-wide' and its mission is to 'be the international authority and advocate on design management'.

Design registration A means of design protection.

Design Research Society (DRS) An international organisation that was established, in recognition that design was common to many disciplines, to explore socially responsible, explicit, and reliable design procedures.

Design right Legislation to protect original design for goods, products and packages against copying without the need to prove a breach of copyright.

Designed Made specifically to some kind of plan.

Desk research Where information is gleaned from work already carried out by other individuals or groups.

Devoré See Burn-out prints.

Dial The needles arranged horizontally inside the cylinder of a circular knitting machine; equivalent to the flat machine's back bed.

Direct costs Those costs that can be directly attributed to a product or service. They can include raw materials, labour costs for weaving or knitting, any finishing cost (per metre, for example, for treating a fabric for flame retardancy) and distribution costs.

Disperse dyes Dyestuffs which are insoluble in water but are applied as a fine suspension or 'dispersion'.

Dobby shedding A mechanism attached to a loom for controlling the movement of the shafts through which the warp yarns are threaded.

Domestic furnishings Furnishings found in the home.

Draft The instruction as to how the threads in the warp are drawn through the healds on the shafts.

Drawthreads In knitting, threads that are introduced that, on removal, allow separation of pieces of fabric with welt edges.

Engineered designs Those where the design is worked to fit the shape of the intended product. Examples of engineered designs include towels, rugs, duvets and headscarves.

ESMA Menswear show in Milan.

Ethnic designs Traditional type designs of non-western origin.

Expofil Yarn show in Paris.

Fabrex Fabric show in London.

Fabric specifications/making particulars Full instructions as to how to make a fabric.

Fancy yarns Yarns developed with deliberately introduced periodic effects.

Fashion adoption The take-up of new styles.

Fashion colours Colours within a colour range perceived as being fashionable or 'of that moment'.

Fee letter A letter from the designer to the client confirming clearly the brief as understood by him, just what is going to be done and by when, what the fees are going to be and how these will be charged.

Felt Fabric made by matting animal fibres together.

Fibres Materials characterised by high ratio of length to thickness. Fibres can be natural or man-made.

Field research Information found directly — by going out and looking; by asking questions to relevant individuals and groups; by using mailed questionnaires, setting up focus groups and interviewing people.

Finish A process or substance applied to enhance the appearance (such as brushing) or performance (such as flame retardancy).

Fixed fees Fees for a project that do not alter.

Flat knitting Knitting produced on machines that have the needles arranged in a straight line on a needle bed, with knitting taking place from side to side, the resulting fabric being produced in open width.

Flocking The application of short fibres to a base fabric by the direct printing of adhesive onto the fabric in the desired areas and then sticking the fibres to these areas.

Focus groups Groups of customers/potential customers brought together by market researchers to find out likes/dislikes, etc.

Forecasting Predicting fashion trends.

Forecasting services Publishers and other organisations offering fashion prediction advice.

Formal balance Where the design elements are almost equally distributed.

Foulards Small patterns repeated directly under and across from one another at measured intervals. The name comes from the French word for scarf or necktie.

Freelance designers Self-employed designers who may work independently, through an agent or through a studio.

Fully-fashioned knitwear Knitwear made from garment pieces shaped (fashioned) on the knitting machine to the correct size and shape required.

Furnishings Product group including curtains, upholstery fabrics, carpets and wall coverings.

Gantt chart Simple horizontal bar chart that graphically displays the time relationships of the stages in a project. Each step in the project is represented by a line or block placed on the chart in the time period in which it is to be undertaken. When completed, a Gantt chart shows the flow of activities in a sequence, as well as those that can be under way at the same time.

Garment blank Panel of fabric knitted for cut and sewn knitwear production.

Geotextile A textile used in soil and soil-based applications such as road building, dams and erosion control.

Grey cloth Undyed and unfinished fabric straight from the loom.

Grosgrain A fine plain weave fabric with weftways ribs, with the warp usually made from continuous filament yarn.

Hand-painted mordanted cottons Cotton fabrics coloured by patterning with mordants and dyeing.

Handkerchief designs Engineered prints to fit into a square of fabric.

Handle (hand in USA) The way a fabric feels.

Haute couture Literally 'high fashion'.

Heald Wires on wooden frames or shafts, with eyes in the middle through which warp yarns are threaded.

Heimtex Furnishing show in Frankfurt.

Horizontal organisations Companies (often small but not always) specialising in only one production process such as knitting, printing, dyeing and finishing.

Hosiery Can mean all types of knitted fabrics, and goods made from knitted fabrics — or stockings, tights and socks.

Household textiles Textile products used within the home, except furnishings. Included are sheets, pillowcases, towels, blankets, and table cloths.

Hue The 'colour' of a colour, e.g. its redness, greenness or yellowness.

International Colour Authority (ICA) Colour forecasting publishers, publishing forecasts twice a year, 24 months ahead of season.

International Council of Societies of Industrial Design (ICSID) An international, non-profit making, non-governmental organisation established to advance the discipline of industrial design.

Igedo Fashion fair in Dusseldorf.

Indirect costs Those costs associated with manufacture that cannot be directly attributed to any individual product or job. Such costs are also termed overheads.

Industrial textiles Textile product group that includes car tyres, medical textiles, geo-textiles, filters, conveyor belts, car safety belts and parachute cords.

Infamoda Inc. Fashion and trend forecasting agency.

In-house/staff designers Designers employed by a company, usually on a full-time basis (although some may be employed part-time).

Institutional fabrics Household textiles used in the contract market.

Interlock fabrics Weft-knitted fabrics that are intermeshed so that the stitches 'interlock' together in such a manner that unroving the resultant fabric is more difficult.

Interstoff Fabric show in Frankfurt.

Island designs Prints where the pattern units making up the designs are surrounded by substantial portions of ground.

Jacquards Mechanical devices that control the patterning of large design areas.

Jersey A generic term for knitted piece goods: continuous weft-knitted fabric.

Knit fabric structure The way the threads interlace.

Knitted fabrics Fabrics produced by interlacing loops of yarn.

Knitting The fabric production process whereby the fabric is formed by intermeshing loops of a single yarn or set of yarns together.

Knitting Industries' Federation (KIF) The National Employers' Organisation for the UK knitting industries.

Knitwear Knitted garments, usually with integral ribs.

Knotting A process used for macramé, fishing nets and some rugs and carpets.

Lace An open-work fabric made by looping, plaiting or twisting threads by means of a needle or a set of bobbins.

Layouts The arrangement of motifs in the framework of the design plane.

Licence Permission, usually in writing.

Lifting plan The instructions for the lifting and lowering of the shafts.

Linking A sewing process used to attach knitted trimmings to body pieces by chain-stitching individual stitches to the body.

Loom A machine for producing cloth by weaving.

Machine gauge The number of needles on a knitting machine per unit of length (usually this unit of length is an inch).

Macramé A type of knotted construction.

Making particulars The full details of manufacture.

Man-made fibres Fibres that are either regenerated or synthetic; viscose rayon, based on regenerated cellulose, is man-made but not synthetic while polyester, polypropylene and nylon are all synthetic fibres. Cellulose acetate may be regarded as semi-synthetic.

Marketing The business of selling goods, including advertising and packaging.

Merchandise Goods for sale.

Merchandising Having the right goods of the right quality at the right time, in the right place, in the right quantity, at the right price.

Merino Breed of sheep and wool from this breed.

Mill Name for factory producing textiles and textile products.

Mood/theme boards Visual presentations of ideas to illustrate a mood or theme.

Motif Recurring element in a design.

Natural dyes Dyes obtained from plant, animal and mineral sources.

Natural fibres Include animal fibres (e.g. wool and silk), vegetable fibres (e.g. jute and cotton) and mineral fibres (e.g. asbestos).

Network analysis A generic term used for several project planning methods, of which the best known are PERT (Programme Evaluation and Review Technique) and CPA (Critical Path Analysis).

Non-wovens Fabrics made from webs of fibres held together and then bonded by some method.

Notation systems Ways of graphically representing woven and knitted fabric structures.

Ogees Symmetrical onion shapes.

Overdyeing When dyeing takes place on top of a previous colouring process.

Overlocking A sewing process which cuts and forms stitches round a raw edge, preventing the fabric from unravelling.

Paisley pattern Distinctive tear-drop shape decorative pattern based on traditional Indian cone or pine motifs.

Pantone Colour referencing system.

Paperwork Designers' initial drawings and design ideas produced on paper.

Passing off Intentionally trying to deceive consumers into thinking a design is by a particular individual or company. This happens frequently with designer labels and branded textiles.

Patent Sole right of manufacture given by Government to originator of novel product or process.

Pattern co-ordination Co-ordination usually achieved by using selected shapes and motifs from the main design to produce related patterns.

Pattern direction The way a pattern lies on a fabric.

Peg-plan The method of instructing the dobby mechanism on a loom to lift the required shafts.

Perceived value The value that a product is seen to have.

Percentage composition Fibre content expressed as percentages.

Performance requirements How a fabric or product has to function.

Picking order The smallest number of picks in colour and/or count that repeats up and down the fabric.

Picks The weft threads that run across the cloth, working under and over the warp ends from selvedge to selvedge.

Pictorial and figurative designs Designs featuring visual representations of objects, people and animals.

Piece A length of woven or knitted fabric.

Piece-dyed Fabric dyed in fabric form rather than, e.g., made from dyed yarn.

Pigments Insoluble colours which are fixed by a binder that bonds them to the fabric.

Pitti Filati Yarn show in Florence.

Plain-knit fabric Single jersey fabric with all face loops on one side and all reverse loops on the other.

Plain weave The simplest weave, with maximum interlacing points.

Planning Managing and controlling events to achieve a goal and making the best use of resources.

Point paper A special graph paper for designing woven fabrics. The standard point paper used is ruled in groups of 8×8, separated by thicker bar lines.

Poplin A fine plain weave cloth with fine weftways ribs, usually made in cotton.

Post-purchase feelings How a purchaser feels after buying something.

Première Vision Fabric show in Paris.

Prêt-à-porter Garment show in Paris.

Price points Prices at which products will retail.

Primary research See Field research.

Print producers Designers putting together ranges of printed textiles.

Printing Methods of colouring some areas of fabrics differently from others by using dyes, pigments and/or paints.

Professionalism Demonstrated by self-confidence and flair, capability and expertise, rational and systematic thought, creativity and judgement, sensitivity to the environment, and to other professions, nationalities and cultures, appreciation of design within a global context, and a lifetime commitment to personal education and professional advancement.

Profit The monetary gain from being in business; the excess of revenues over outlays and expenses in a business enterprise.

Promostyle Trend and colour forecast publisher and consultancy.

Purchase invoices Invoices of items bought.

Purl fabric Knitted fabric with face and reverse loops in the same wale.

Radial balance This is when design elements radiate from a central point, as the spokes of a wheel or in the natural form of a daisy.

Range A group of fabrics (or products) designed, developed and edited to be shown and sold to the market each season.

Range planning Managing and controlling events to put together a range.

Reactive dyes Those dyes which combine chemically with the fibres of a fabric when fixed.

Reed What the warp yarns are threaded through to keep the threads spaced correctly during weaving. The reed also often beats up the picks as the cloth is woven.

Repeat The smallest size of a pattern that shows the full pattern.

Repeat artists Designers who take designs and put these into a size and repeat appropriate to the intended end-use.

Research Finding information; often about products and markets, production processes and techniques, etc.

Resist A substance that will not take up colour.

Retail The sale of goods in small quantities to consumers.

Retail selling seasons Those times of the year when Spring/Summer and Autumn/Winter product ranges are sold.

Retainer Fee paid in advance to ensure services of a designer.

Rib fabrics Very elastic knitted structures with good recovery, knitted on machines with two sets of needles.

Rib gaiting The needle arrangement on rib machines where the needles are arranged in such a way as to allow them to intermesh when raised.

Roller printing A method of printing fabric that uses engraved copper rollers.

Royalties Payments made to a designer as a percentage of revenue from sales.

Sales invoices Invoices for goods sold.

Sample blankets/section blankets Woven trial pieces with differing sections across the warp. These sections may vary in colour, yarn type and/or weave.

Satin and sateen Weaves that result in fabrics of a smooth and lustrous appearance.

Screen printing A method of printing whereby the colour is applied by what is essentially a stencilling process.

Seasonal ranges Ranges produced for the selling seasons Spring/Summer and Autumn/Winter. Mid-season ranges may also be produced.

Secondary research See Desk research.

Section blankets See Sample blankets.

Selling price The price the product is sold at.

Sett In a woven fabric, the number of warp threads or 'ends' used per inch/centimetre and the relationship of these to the number of weft threads or 'picks' per inch/centimetre.

Source To find suppliers of required materials.

Spinners Manufacturers of yarn.

Staff designer See In-house designer.

Stentering A controlled straightening and stretching process.

Stitch bonding A method of stitching webs of fibres together, resulting in a non-woven fabric.

Stolling Rib trim put on so that the rib is at right angles to the direction of the rib on the body.

Strapping Narrow width rib trim, usually 1×1 rib or of a half milano construction, applied so that the rib runs along the garment lengthways.

Strike-offs Sample prints on fabric for pattern and colour approval.

Studio The place where a designer works. It can be anything from an area set aside on the factory floor to a large smart office, or even a suite of offices.

Stylist Designer who puts together a range of products.

Subtractive colour mixing The subtractive primary colours are cyan, yellow and magenta and when mixed together they subtract from the light producing black. When different pairs of the subtractive primaries are mixed, the colours red, green and blue are produced.

Synthetic dyes Dyes synthesised from organic molecules.

Taffeta A plain weave fabric characterised by indistinct weftways ribs.

Tapestry weave A weft-faced plain weave where the weft threads are packed closely together so that the warp is hardly seen. Tapestry is often associated with the large pictorial wall hangings of medieval and later Europe. However, strictly speaking, tapestry is a distinctive woven structure—a weft-faced plain weave with discontinuous wefts.

Textile Strictly speaking the definition of the word textile is 'a woven fabric'. However, the term textile is now considered to cover any product that uses textile materials and is made by textile processes.

Textile Institute Worldwide professional association for people working with fibres and fabrics, clothing and footwear, interior and technical textiles.

Tie dye A method of patterning fabric by tying areas of fabric and then dyeing.

Tile (or brick) repeat A simple repeat where the motifs are repeated rather like a simple brick wall pattern. The second row slides halfway across, in a widthways direction.

Time management Making the best use of time to achieve what is necessary.

Tissu Premier Fabric show in Lille.

Toiles de Jouy Fabrics in a pictorial style named after copper-plate printed patterns produced by the Jouy factory in France.

Tossed patterns An arrangement of motifs where the motifs in a repeat do not occur at regular intervals. It is as if the designs are dropped in a fairly random way onto the fabric.

Trade and service mark Officially registered names or symbols used to distinguish a product.

Trade exhibitions Exhibitions/shows held to promote specific ranges of products.

Trade organisations and associations Organisations associated with particular business areas.

Transfer (or sublimation) printing A method of printing that uses pre-printed papers to transfer the design to the fabric.

Trend report A report on general tendencies and directions.

Tricks In knitting, the slots which space the needles.

Trimmings Pieces used to decorate or complete products.

True designs In weaving, those parts of a section blanket where a weft section is woven on the warp section for which it was intended.

Twelfth/ 1/12 Single garment from a production batch of 12.

Twill weaves Woven fabrics with diagonal lines formed by the weave structure.

Unbalanced colourways When there are no similar colour relationships between colourways, giving the colourways a different appearance to the original design.

Unroved Unravelled.

Value When applied to colour, the amount of lightness or darkness.

Vertical organisations Mills that take in fibre, spin yarns, dye these and then either weave or knit these into fabrics/garments.

Voile Lightweight, open, plain weave fabric.

Wales Vertical columns of knitted stitches.

Warp The lengthways threads in a woven fabric.

Warp knitting A method of knitted fabric construction whereby a set of warp yarns are simultaneously formed into loops, connected by a sideways movement that is such as to cause the loops to interlink.

Warping order The smallest number of ends in colour and/or count that repeats across the fabric.

Weave repeat The smallest number of different intersections between warp and weft that, when repeated in either direction, gives the weave of the whole fabric.

Weft The widthways threads in a woven fabric.

Weft knitting A method of knitted fabric construction whereby loops are formed, one at a time, in a weftways direction as the fabric is formed.

Welt In knitting, a secure edge.

Wholesalers Essentially, businesses that buy from a manufacturer and, without changing the product, sell it in smaller quantities to retailers or smaller manufacturers.

Woollen Description of wool yarns, or fabrics made from this yarn, which is produced by carding, condensing and spinning.

Woolmark Company Organisation that exists to increase demand for wool, world-wide.

Worsted Description of smooth woollen yarn, spun from staple fibre, which is combed for smoothness, and the fabric made from this yarn.

Woven fabrics Fabrics that consist of two sets of threads, the warp and weft, which are interlaced at right angles to each other.

Yarn producers See spinners.

Yarn production Essentially about taking fibres, organising them so that they lie in a lengthways direction and twisting them to create a yarn.

Yarns Constructions of substantial length and relatively small cross-section made from fibres, usually with some twist.

Index